science

The Salters' Approach
Key Stage 4 Book 1

Bob Campbell John Lazonby Robin Millar Steve Smyth

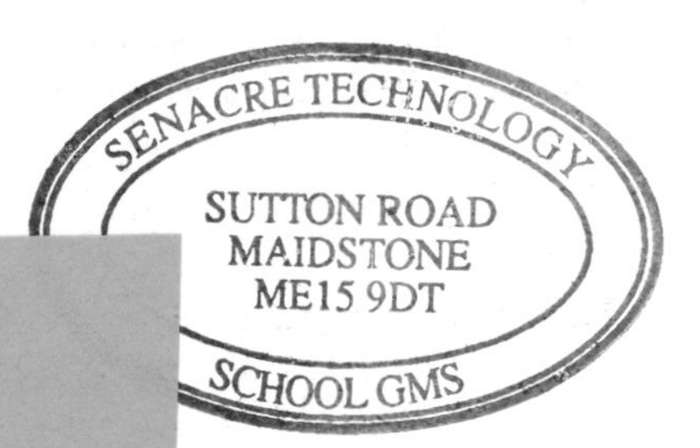

HEINEMANN
EDUCATIONAL

The authors and publishers are grateful to the following for permission to reproduce photographs:

Contents page:
TL Trevor Hill; BL Shell UK; R Sporting Pictures UK Ltd.; TL Robert Harding Picture Library; BL Trevor Hill; R J.C. Allen/Frank Lane Picture Agency.
Energy Matters:
p4 Trevor Hill, T Michael Prior; p5 T Robert Harding; M Michael Prior; p8 TR Sporting Pictures UK Ltd.; BR Michael Prior; BL Robert Harding; ML Vauxhall Motors; MR Trevor Hill, p9 ALL Trevor Hill; p10 TR Valor Gas, BL Michael Prior; ML British Gas; MR Trevor Hill; p11 TL Trevor Hill; BL Michael Prior; BR Robert Harding; p13 Philips Lighting; p14 Michael Prior.

Keeping Healthy:
p15 L Sporting Pictures UK Ltd.; TR Grapes/Michand/Science Photo Library; MR CNRI/Science Photo Library; BR St Bartholomews Hospital; p17 Biophoto Associates/Science Photo Library; p18 T St Bartholomews Hospital; B Hulton Picture Company; p19 Wellcome Institute Library, London; p20 TL Science Photo Library; TR Robert Harding; BR Science Photo Library; p21 St Bartholomews Hospital; p22 Trevor Hill; p24 John Olive/Chemistry Dept, University of York; p25 John Olive/Chemistry Dept, University of York.

Transporting Chemicals:
p31 TL Shell UK; BL Hugh Ballantyne/Millbrook House; TR Shell UK; BR Whitbread Brewery; p32 ICI; p33 ICI; p34 TR Vivien Fifield; TR The British Library; B Vivien Fifield; p35 Vivien Fifield; p39 London Fire & Civil Defence Authority; p41 TL Trevor Hill; TR Barnaby's Picture Library.

Construction Materials:
p46-7 Pilkingtons; p48 TL J Allan Cash; TR Elizabeth Ann Kitchens; MR J Allan Cash; ML Armitage Shanks; BR Robert Harding Picture Library; p49 Barnaby's Picture Library; p50 TL Robert Harding Picture Library; M&R John Lazonby; BL Glasheen/Barnaby's Picture Library; M&R John Lazonby; p52 Barnaby's Picture Library; p54 Barnaby's Picture Library; p56 Barnaby's Picture Library; p57 Barnaby's Picture Library.

Moving On:
p59 T&ML Trevor Hill; BL John Lazonby; R Barnaby's Picture Library; p60 TL Barnaby's Picture Library; TR MIRA; BL & p61 Reproduced from *Verkeer and Veiligheid* , PLON, Pub. NIB of Zeist, Netherlands; p64 B&C Trevor Hill; L John Olive; TL The Guardian; p66 Sporting Pictures (UK) Ltd; BR Aviemore Photographic; p70 Trevor Hill; p71 Trevor Hill & B Allan Cash; p73 Reproduced from *Verkeer*, PLON, Pub. NIB of Zeist, Netherlands; p74 T Barnaby's Picture Library; Ford Motor Company; BMW; Sealink; Barnaby's Picture Library.

Food for Thought:
p75 TL L West/Frank Lane Picture Agency; BL ICI; TR Derek Robinson/Frank Lane Picture Agency; BR Hutchison Library; p76-7 S McCutcheon/Frank Lane Picture Agency; p79 L National Dairy Council; TR & BR AFRC Institute of Food Research; M Hellman's; p81 T Gamma Press/Frank Spooner; M John Wright/Hutchison Library, B Trevor Hill; p82 L ICI; TR J Burgess/Science Photo Library; BR H Binz/Frank Lane Picture Agency; p83 L Vivien Fifield; R ICI; p84 Trevor Hill, p86 L Mazola; TR JC Allen/Frank Lane Picture Agency; BR Trevor Hill; p87 Trevor Hill.

(T = Top; B = Bottom; R = Right; L = Left; M = Middle)

Heinemann Educational, a division of

Heinemann Educational Books Ltd
Halley Court, Jordan Hill, Oxford OX2 8EJ

OXFORD LONDON EDINBURGH MADRID
ATHENS BOLOGNA PARIS MELBOURNE
SYDNEY AUCKLAND SINGAPORE TOKYO
IBADAN NAIROBI HARARE GABORONE
PORTSMOUTH NH (USA)

ISBN 0 435 63004 0

First published 1990

92 93 94 95 11 10 9 8 7 6 5 4 3

Designed and typeset by KAG Design Ltd

Printed in Spain by Mateu Cromo

About this book

Salters' Science 1 is different from conventional textbooks, so read this section carefully to find out how to use it. (It's not just a case of reading each chapter from beginning to end!)

This textbook accompanies the first six units of the Salters' Course at Key Stage 4. These six units are:

- Energy matters
- Keeping healthy
- Transporting chemicals
- Construction materials
- Moving on
- Food for thought

Each unit has a corresponding chapter in this book.

Each chapter is divided into five key sections.

- Introducing
- Looking At
- In Brief
- Thinking About
- Things to do

These sections are meant to be used in different ways. Have a quick look at a chapter to see what each section is like and then turn back to this page and read about how to use them.

Introducing

This page sets the scene and tells you why the topic is important and what the chapter covers.

You could read this page before you start to study the topic in class.

Looking at

These are several Looking At pages in each chapter. Each page uses coloured photographs and diagrams to show an important application or use of the science ideas. The page sets you tasks to help you to understand these ideas.

Your teacher will ask you to work through some Looking At pages and to do the tasks on the page.

In brief

This section presents a summary of what you need to know and understand about the topic.

You could use the In Brief when you have finished a unit or when you are revising. Key ideas from this section are explained more fully in the next section.

Thinking about

This explains the key scientific ideas developed in the unit.

When revising, you might find a part of the In Brief that you need to do more work on. You can then move to the Thinking About to find out more about it. Your teacher might ask you to read parts of this section after you have met the ideas in class.

Things to do

This is a collection of things for you to do. There are:

- Activities to try
- Things to find out
- Things to write about
- Points to discuss
- Questions to answer

You could use a selection of these either in class or for homework.

Contents

ENERGY MATTERS 1

KEEPING HEALTHY 15

TRANSPORTING CHEMICALS 31

CONSTRUCTION MATERIALS 45

MOVING ON 59

FOOD FOR THOUGHT 75

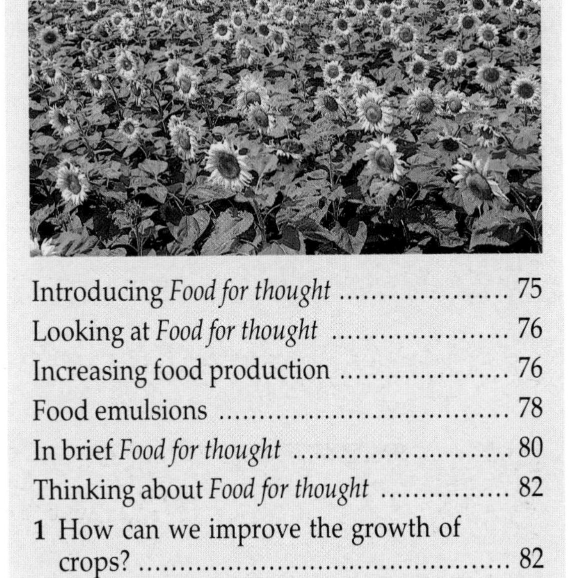

Introducing

ENERGY MATTERS

How often do you stop and think about 'energy matters'? Probably not very often. How important is it for you to care about energy? The picture on this page shows a large number of appliances like toasters, fridges and so on which most definitely affect your life. These appliances use concentrated sources of energy to do jobs for you. You yourself use another source of concentrated energy about three times a day, just to keep you going.

1 Make a list of all the appliances shown which use energy.
2 Try to classify the appliances. For example, you could put them in groups according to what energy source they use, or which you think use a lot of energy and which use hardly any.
3 Which appliances do you think might have required a lot of energy to make them?
4 Which appliances do you think might not use energy very efficiently?

IN THIS CHAPTER YOU WILL FIND OUT

- how to measure or calculate the amount of energy used by different appliances or devices
- how to calculate the cost of the energy you use
- what happens to the energy you use
- that energy is also needed to make things
- how we can use energy more efficiently.

Looking at

Cutting Heating Costs

Each year we use a vast amount of energy to heat our homes. The cost of this energy in 1989 was about £10 billion in the U.K. How can we reduce this?

Some of the energy used in heating homes is wasted. One way of cutting waste is to use insulation. Of course, insulation itself costs money to install. The diagram shows the different kinds available, how much they cost and how much they save.

1. Use the information to draw up a chart showing the cost of each type of insulation, and the saving you would expect to make each year.
2. Work out how long it would take to recover the cost of each type of insulation from the savings made. This is called the *pay-back time.*
3. Make a list of types of insulation in order of cost effectiveness.
4. Find out how thick loft insulation should be.

HOT WATER CYLINDER JACKET

It costs £10 to fit an insulating jacket around a hot water tank. The saving is about £60 a year.

LOFT INSULATION

A mineral fibre blanket can be used to cover the floor in the loft. This cuts down energy loss through the roof by 80% and saves about £100 a year. It costs about £200 to install.

DRAUGHT EXCLUDERS

Covering gaps under doors and windows can be done cheaply – £60 on this particular house with an energy saving of about £15 a year.

DOUBLE GLAZING

Double glazing is expensive to install (about £1500 per average house) and saves about £50 a year. It has other advantages though, such as cutting down noise from outside.

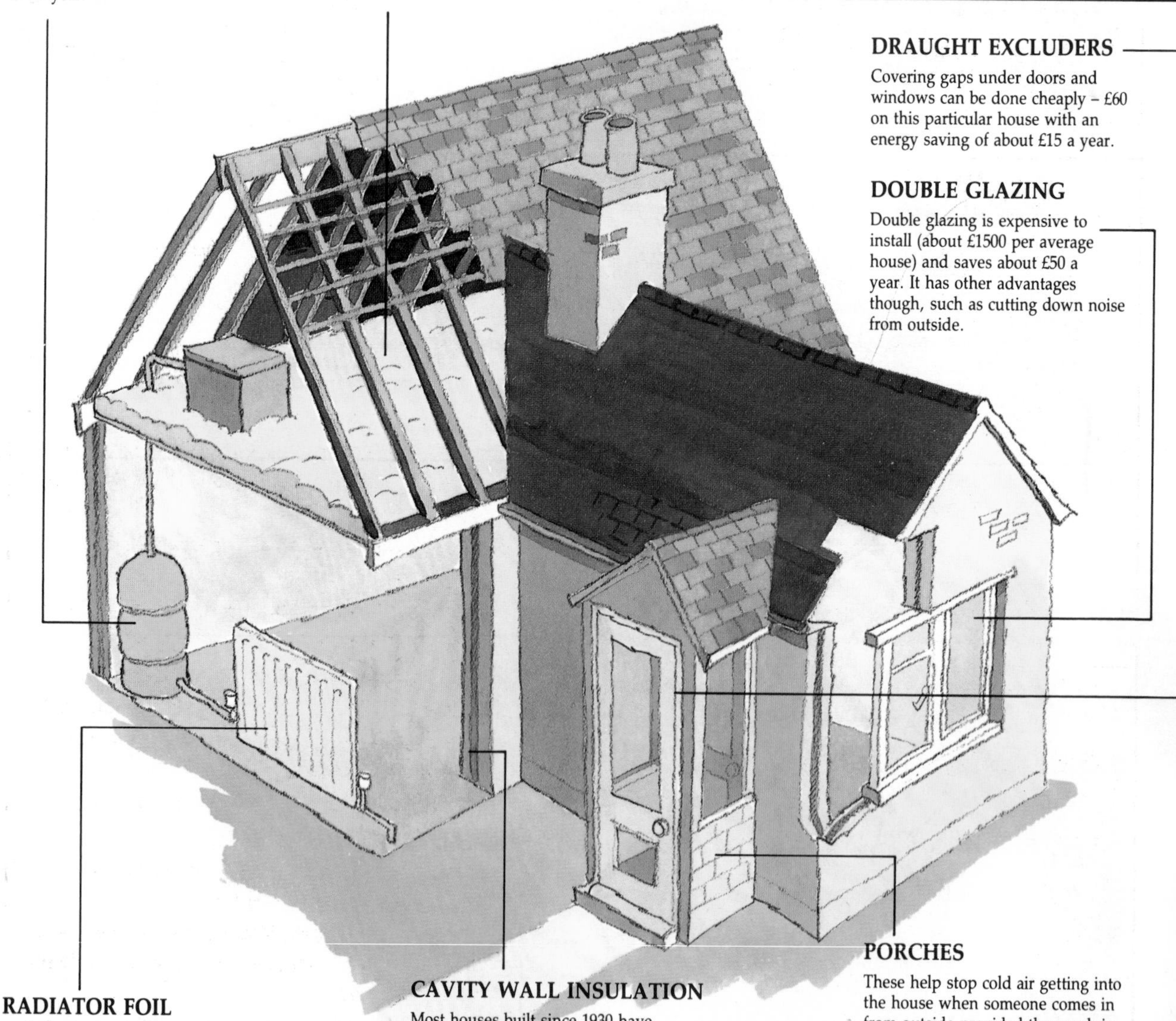

RADIATOR FOIL

Fitting foil behind radiators on external walls reflects back energy that would otherwise be lost. It probably only saves £10 a year, but the cost of the foil is just £2.

CAVITY WALL INSULATION

Most houses built since 1930 have two walls with a gap between them. Filling this gap with foam will cost about £500. Less energy will be lost through the walls, saving about £100 a year.

PORCHES

These help stop cold air getting into the house when someone comes in from outside provided the porch is big enough for the outer door to be shut before the inner one is opened. A porch for this house would cost £1000, and save £40 a year in heating costs.

Looking at

Where the Energy Goes

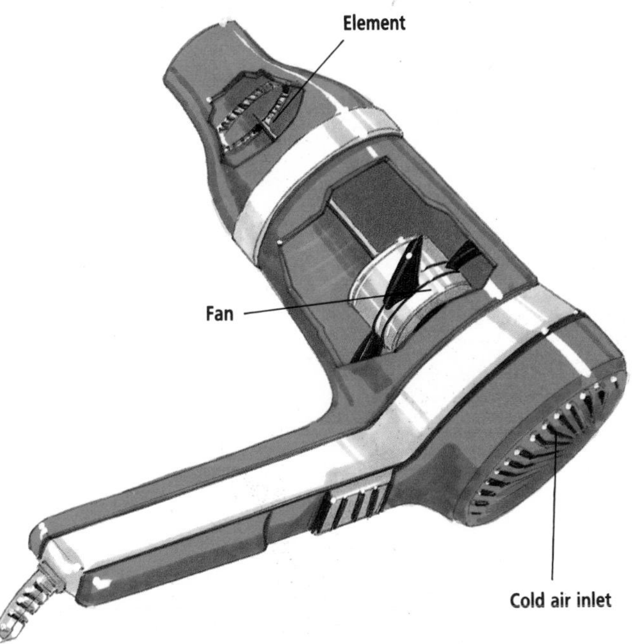

1. Draw an arrow diagram (see page 10) to show the energy changes in a hairdryer when it's working properly.
2. Is the hairdryer using the same amount of energy when its fan stops working?
3. Write a sentence to explain each of the following. In each sentence use one of these words:
   conduction, convection, radiation.
   - how the handle of the hairdryer becomes hot
   - how the energy spreads from the glowing element to Ravinda's hair
   - how the air above it also gets hot when the dryer is switched on
   - how the energy spreads from the red hot element after the fan stops.
4. Write a sentence to explain why the element glows red hot when the fan stops, but doesn't while the fan is working.
5. What energy changes happen when someone dries their hair with a towel?

Looking at

A Clear Saving

Glass is a very useful material when it comes to making containers and bottles. It is strong and very resistant to corrosion. As it is clear, you can see what is inside. It can also be very beautiful and is a traditional container for drinks.

Why do you think glass is used for these objects?

The Life of a Bottle

Sand, limestone and salt are quarried. The salt is made into soda ash and all these materials are taken to the glass making factory.

All these materials are processed and heated in a furnace. They melt and combine to form glass.

The glass is shaped, cleaned and sterilised.

The bottles are filled and taken to the shops.

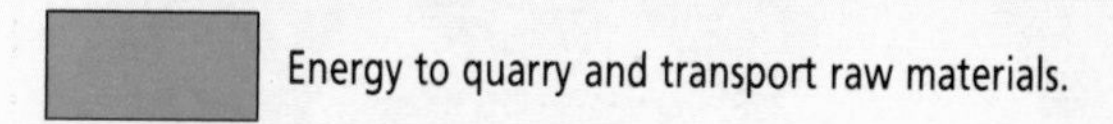

Energy to quarry and transport raw materials.

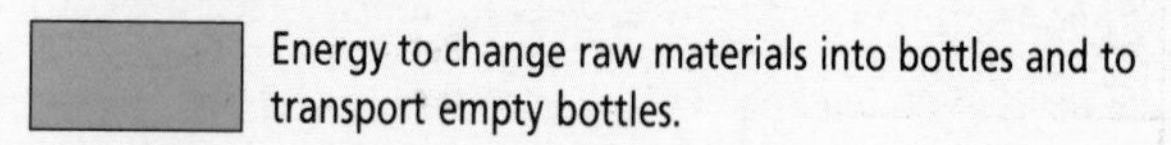

Energy to change raw materials into bottles and to transport empty bottles.

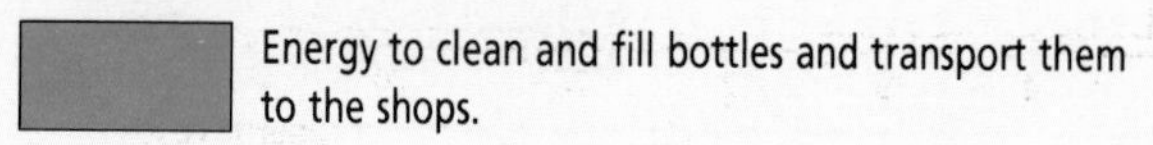

Energy to clean and fill bottles and transport them to the shops.

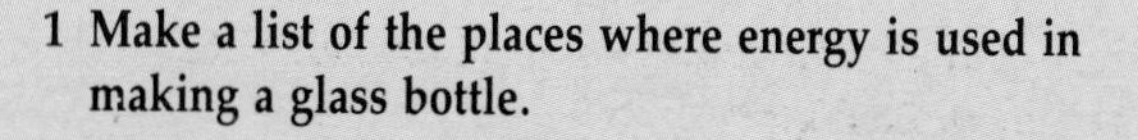

1 Make a list of the places where energy is used in making a glass bottle.

Save energy by recycling . . .

Recycling glass can save a lot of energy. Broken glass is added to the raw materials in the furnace. This means that less energy has to be used to melt the materials and so make the glass. This saves 1650 MJ (that's 30 gallons of heating oil!) for each tonne of glass used. A tonne of glass is about 4000 milk bottles!

2 Look back at the history of a glass bottle. What other major energy costs could be saved by recycling glass, rather than by using raw materials?

Recycling glass – where are the bottle banks near you?

. . . and more by re-using

Even more energy savings can be made by re-using bottles. More energy is used to make each refillable bottle because it has to be stronger and more difficult to break. But this extra energy cost is soon saved if the bottle is re-used as the table below shows. The colours on the table and the key show *how* the energy is used.

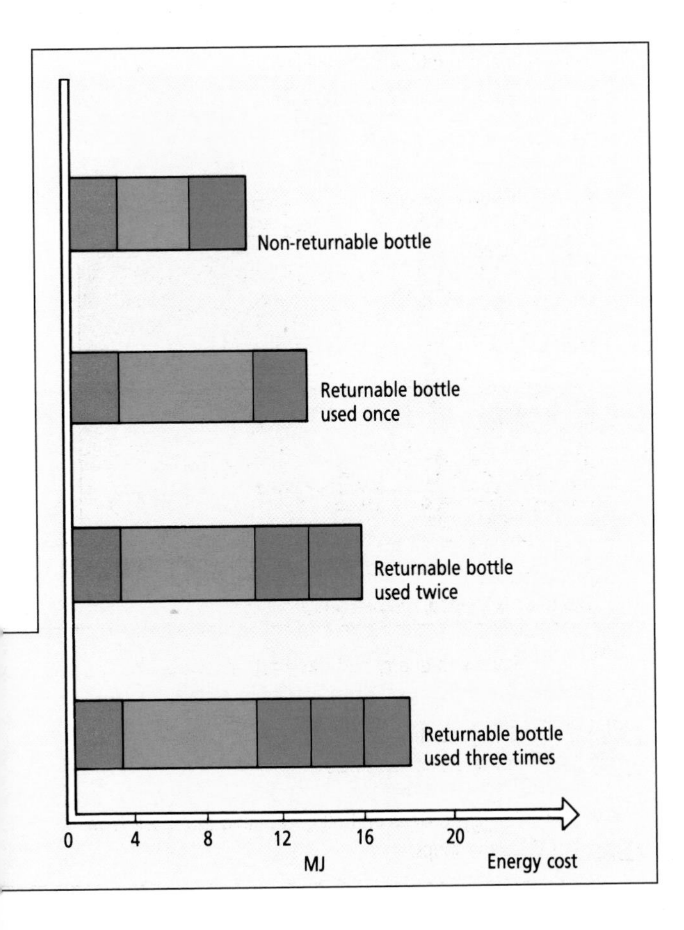

How can you persuade people to re-use bottles?

3 What is the *average* energy cost for a refillable bottle that makes:
(a) 1 trip, (b) 2 trips, (c) 3 trips?

4 Design a poster or a leaflet as part of a campaign to encourage people to re-use bottles rather than recycle broken glass.

Looking at

How Big Should the Bill Be?

These four houses are up for sale. You'll notice that there are a number of similarities and differences between them. In choosing a home, the size of the heating bills will be an important consideration.

1 Decide which of the electricity and gas bills, A to D, probably come from which house. The bills are for the winter quarter.

Start by drawing a table which compares the houses. Include the features of each house which you think could influence the bills.

House No.	Gas c.h.	Insulation

Write a brief explanation for each of your decisions.

1 A well presented 2 year old, 3 bedroomed detached house. Gas central heating, cavity wall insulation, fitted kitchen, burglar alarm, main bedroom with en suite shower room and fitted wardrobes. Garage, gardens.

£85,500

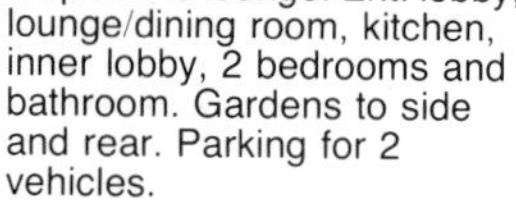

2 A well presented one year old, end of four, 2 bedroomed 'semi'-detached bungalow with excellent features which include Economy 7 electric heaters, feature stone fireplace to lounge. Ent. lobby, lounge/dining room, kitchen, inner lobby, 2 bedrooms and bathroom. Gardens to side and rear. Parking for 2 vehicles.

£52,950

3 A very well presented and surprisingly spacious 2 bedroomed mid terraced house with benefits that include 2 good sized reception rooms, fitted kitchen and first floor modern bathroom. Gas central heating.

£44,500

4
- A sup. individ. des. det. house set in ½ acre
- Solid fuel c.h.
- Porch, hall, lounge, d/room
- Study/bed 4, kitchen, rear hall, utility
- 3 further beds, 1 with en suite, 2 further bathrooms
- Lge dble garage, boiler room, fuel store, ext. w.c.
- Summer house, lge conservatory, sec. system

£190,000

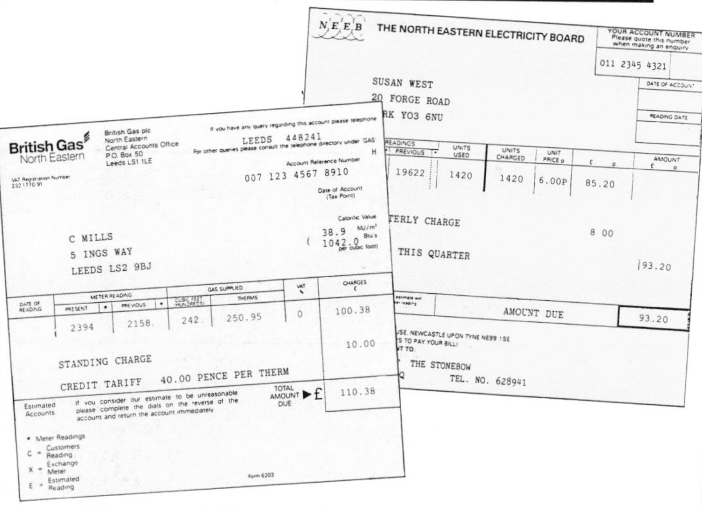

N.E.E.B THE NORTH EASTERN ELECTRICITY BOARD

YOUR ACCOUNT NUMBER Please quote this number when making an enquiry: 011 2345 4321

DATE OF ACCOUNT

READING DATE

SUSAN WEST
20 FORGE ROAD
RK YO3 6NU

READINGS PREVIOUS	UNITS USED	UNITS CHARGED	UNIT PRICE p	£ p	AMOUNT £ p
19622	1420	1420	6.00P	85.20	

TERLY CHARGE 8 00

THIS QUARTER 93.20

AMOUNT DUE 93.20

SE. NEWCASTLE UPON TYNE NE99 1SE
S TO PAY YOUR BILL
T TO:
THE STONEBOW
TEL. NO. 628941

British Gas North Eastern

British Gas plc
North Eastern
Central Accounts Office
P.O. Box 50
Leeds LS1 1LE

VAT Registration Number 232 1770 91

If you have any query regarding this account please telephone LEEDS 448241
For other queries please consult the telephone directory under 'GAS'

Account Reference Number 007 123 4567 8910

Date of Account (Tax Point)

Calorific Value 38.9 MJ/m³ (1042.0 Btu's per cubic foot)

C MILLS
5 INGS WAY
LEEDS LS2 9BJ

DATE OF READING	METER READING PRESENT	*	PREVIOUS	*	GAS SUPPLIED CUBIC FEET (HUNDREDS)	THERMS	VAT	CHARGES £
	2394		2158		242	250.95	0	100.38

STANDING CHARGE 10.00

CREDIT TARIFF 40.00 PENCE PER THERM

TOTAL AMOUNT DUE ▶ £ 110.38

Estimated Accounts: If you consider our estimate to be unreasonable please complete the dials on the reverse of the account and return the account immediately

* Meter Readings
C = Customers Reading
X = Exchange Meter
E = Estimated Reading

Form 6203

These two bills come from two of the houses. And the summary of bills for each house is given in the table.

Bill	Total from bill	
	Electricity	**Gas**
A	£220.91	none
B	£93.20	£180.10
C	£62.76	£110.38
D	£150.12	none

2 Suggest a reason why house 4 uses solid fuel for central heating.

3 For which house would it be easiest to reduce the bills? Explain how you would do this.

4 Why do you think there is a standing gas charge and a quarterly electricity charge?

5 Economy 7 is a cheaper alternative electricity tariff (charge rate). It is used to charge people who have night storage heaters. Suggest why the electricity board has two tariffs.

In brief

Energy Matters

1 Our comfortable lifestyle depends on using a lot of energy, at home and at work.

2 Fuels are concentrated stores of energy.

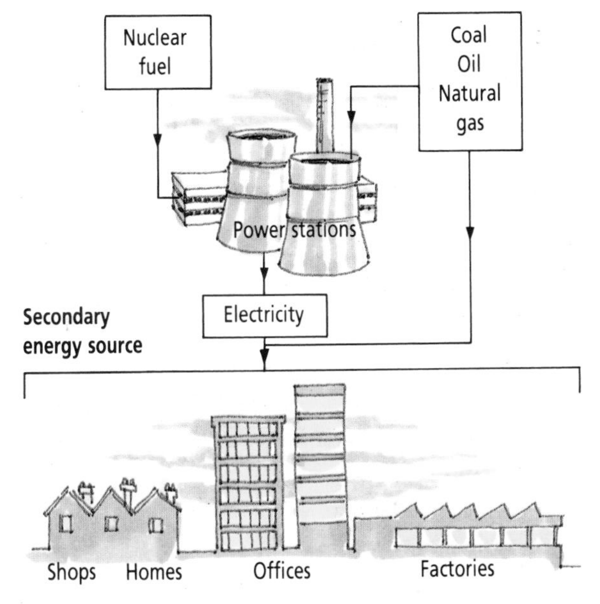

3 Coal, oil and natural gas are fossil fuels. They are the fossilized remains of forests and of small sea creatures which lived millions of years ago. Once we have used them up, they are gone for ever.

4 Different tasks take different amounts of fuel. If we want to measure the amount of energy used, we need a common unit for energy. This is the joule (J).

5 Some electrical appliances use energy more quickly than others. Those which involve heating use a lot of energy.

6 The electricity meter in your home measures how much electrical energy you are using. It 'adds up' the energy used by all your appliances.

7 Every domestic electrical appliance has a power rating in watts (W) written on it. The bigger the power rating, the faster it uses energy. An appliance with a large rating will cost more to run than one with a smaller rating if they are kept switched on for the same time.

8 Appliances do not really 'use' energy – they change it from one form to another, or transfer it from one place to another. This can be illustrated by an energy arrow diagram.

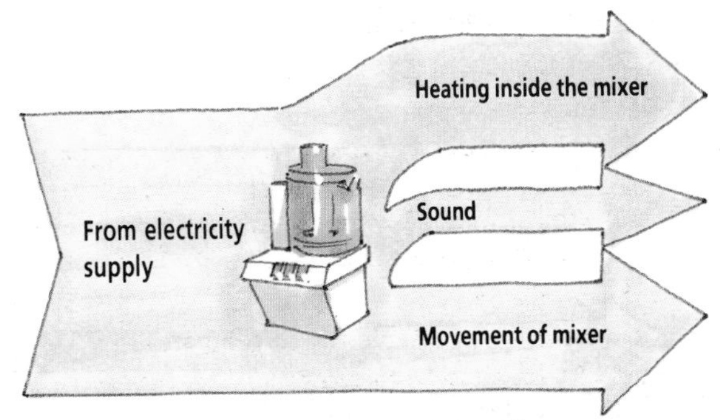

9 There are two important energy *laws*:

- There is always the same total amount of energy after an event as there was at the beginning **(conservation of energy)**.
- Energy always spreads out from concentrated sources (fuels) and ends up in many places. Some always ends up causing unwanted heating **(spreading of energy).**

10 Hot objects cool down and their surroundings get slightly warmer. The energy in the hot object has spread further.

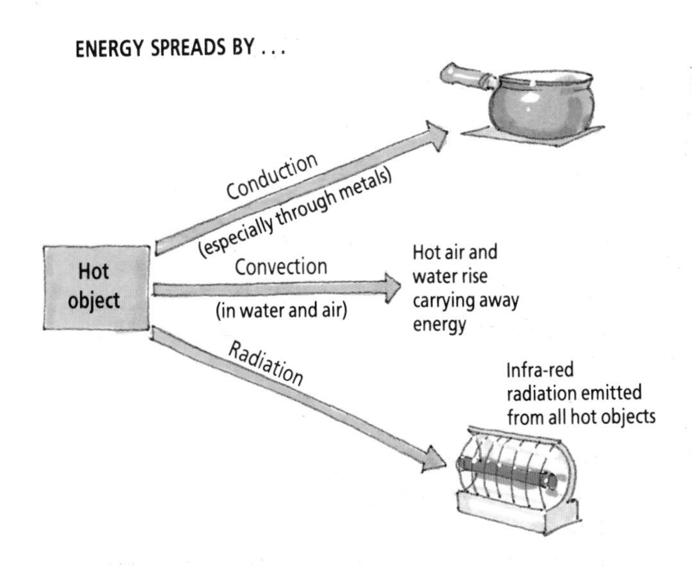

11 We can reduce our fuel bills by insulating our homes better. Insulation makes it harder for energy to spread.

12 Energy is needed to *make* things. Everything has an 'energy cost ' as well as a 'raw materials cost '.

Thinking about

Energy Matters

1. Fuels – making things happen

To make things happen, we always need an energy source.

To run, these athletes are using the energy stored in the food they have eaten.

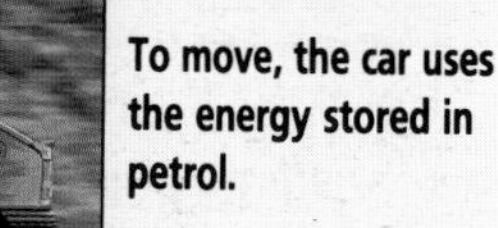

To move, the car uses the energy stored in petrol.

When it is switched on, the torch uses the energy stored in the chemicals in its cells.

To raise the lift and its load, energy is supplied by electricity to the lift motor.

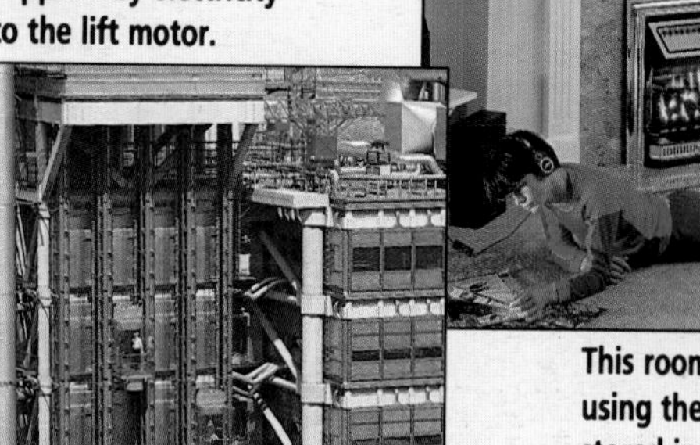

This room is heated using the energy stored in natural gas.

Fuels are concentrated energy stores. Coal, oil, natural gas and wood are **primary fuels.** So are the chemicals inside dry cells (batteries). Electricity is a **secondary fuel.** Mains electricity is generated in power stations by burning coal or oil, or by using another primary fuel – uranium.

Fuels are valuable because they provide concentrated energy which we can use to do all sorts of useful jobs.

2. Measuring energy

(a) Can we measure how much energy we use to do something? One way is to measure how much fuel the task needs. Here is an example:

To make a pot of tea, you have to boil about 0.5 litres of water. At home, you might do this with an electric kettle, or on a gas ring. If you were on a camping holiday, you might do it using a small portable gas stove.

How much fuel does each heating method use?

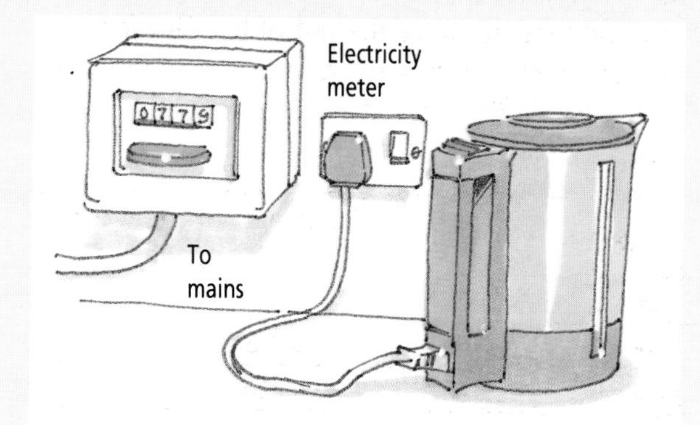

The wheel inside the meter turns 37·5 times before the water boils. The meter is marked '375 revs./kWh'.

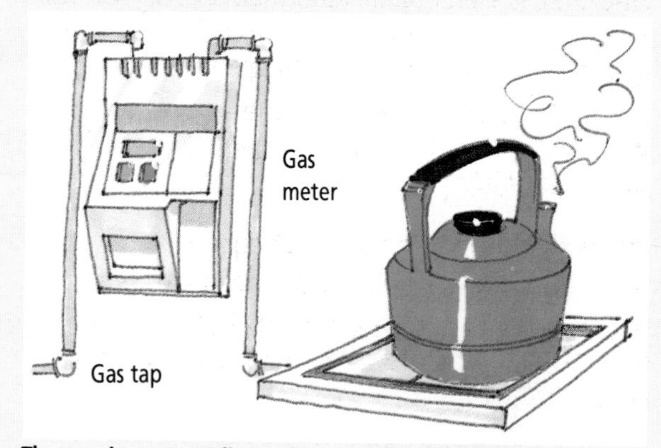

The gas ring uses 18 litres of gas to bring the water to the boil.

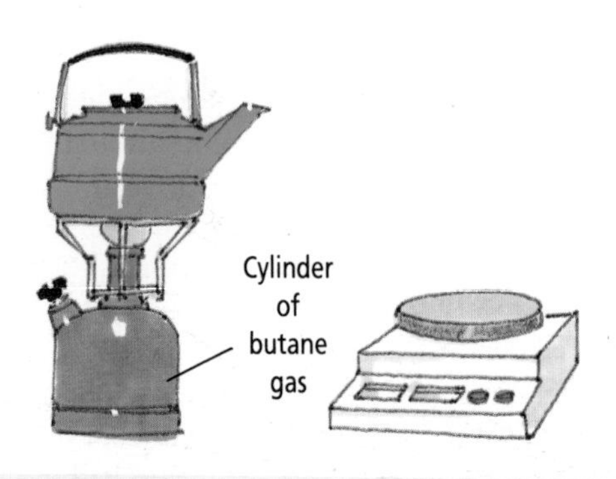

The stove weighs 16 grams less after boiling the water than it did before.

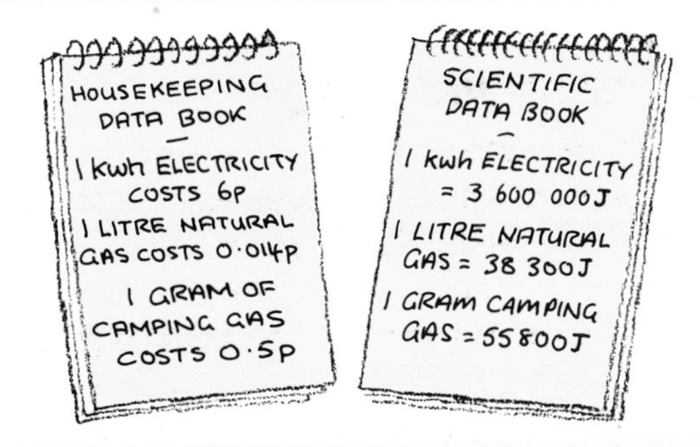

If we know how much we have to pay for each fuel, we can compare the costs of three methods of heating water to make the pot of tea. But for a proper scientific comparison, we need to measure the energy in the same unit each time. The unit used is the joule (J).

Fuel used	Electricity	Natural gas	Camping gas
Amount	0.1 kWh	18 l	16 g
Cost	0.6 p	0.25 p	8 p
Energy	360 000 J	690 000 J	890 000 J

Notice that the numbers of joules are very large. One joule is a very small amount of energy! So it is sometimes more convenient to use the megajoule (MJ). (1 MJ = 1 000 000 J)

Why is the energy used not the same each time? We will come back to that question in *Thinking about 6* on page 12.

(b) Electrical appliances use energy at different rates. Instead of the electric kettle in the diagram, we can connect other appliances to the kilowatt-hour meter and compare how quickly they use electrical energy.

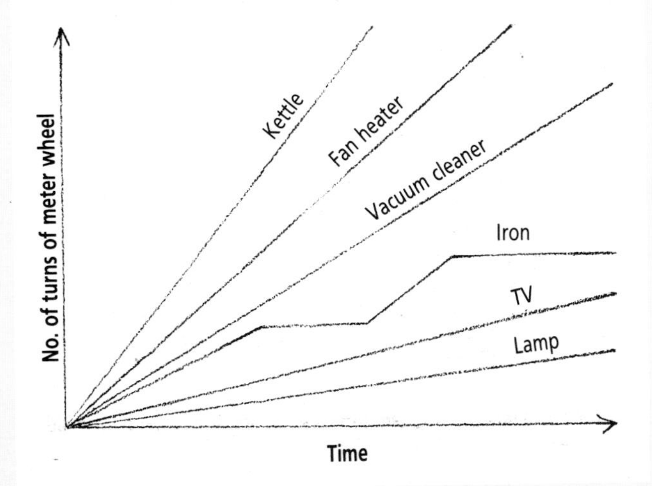

The steeper the graph, the faster the appliance is using energy. Heating is very 'energy greedy'; it uses energy fast! Notice the flat parts of the graph for the iron. An iron has a thermostat. When the iron reaches the temperature it is set for, its heater switches off, even though the iron itself is still switched on. The iron then slowly cools down. At a certain point the thermostat switches the heater back on again. In this way the iron stays at a roughly steady temperature.

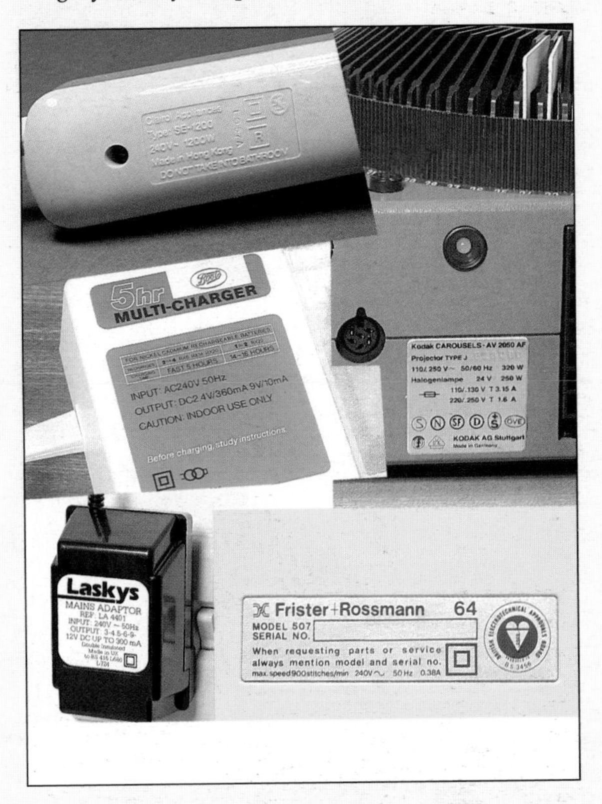

If you want to find out how quickly an appliance uses energy, look at its rating plate. The rating is marked in watts (W) or kilowatts (kW) (1 kW = 1000 W). The bigger the number, the faster the appliance uses energy.

Appliance	Typical rating
Radio	9 W
Lamp	60 W
Television	80 W
Food mixer	300 W
Power drill	315 W
Fan heater	1000 W
Washing machine	1000 W-3000 W
Electric kettle	2000 W
Tumble dryer	2200 W
Immersion heater	2500 W

3. Energy transfers

We often talk about appliances 'using' energy. But what they really do is to transfer energy – from a concentrated fuel to where we want it to go.

An arrow diagram is a good way to picture what is going on from an energy point-of-view. Here are some examples:

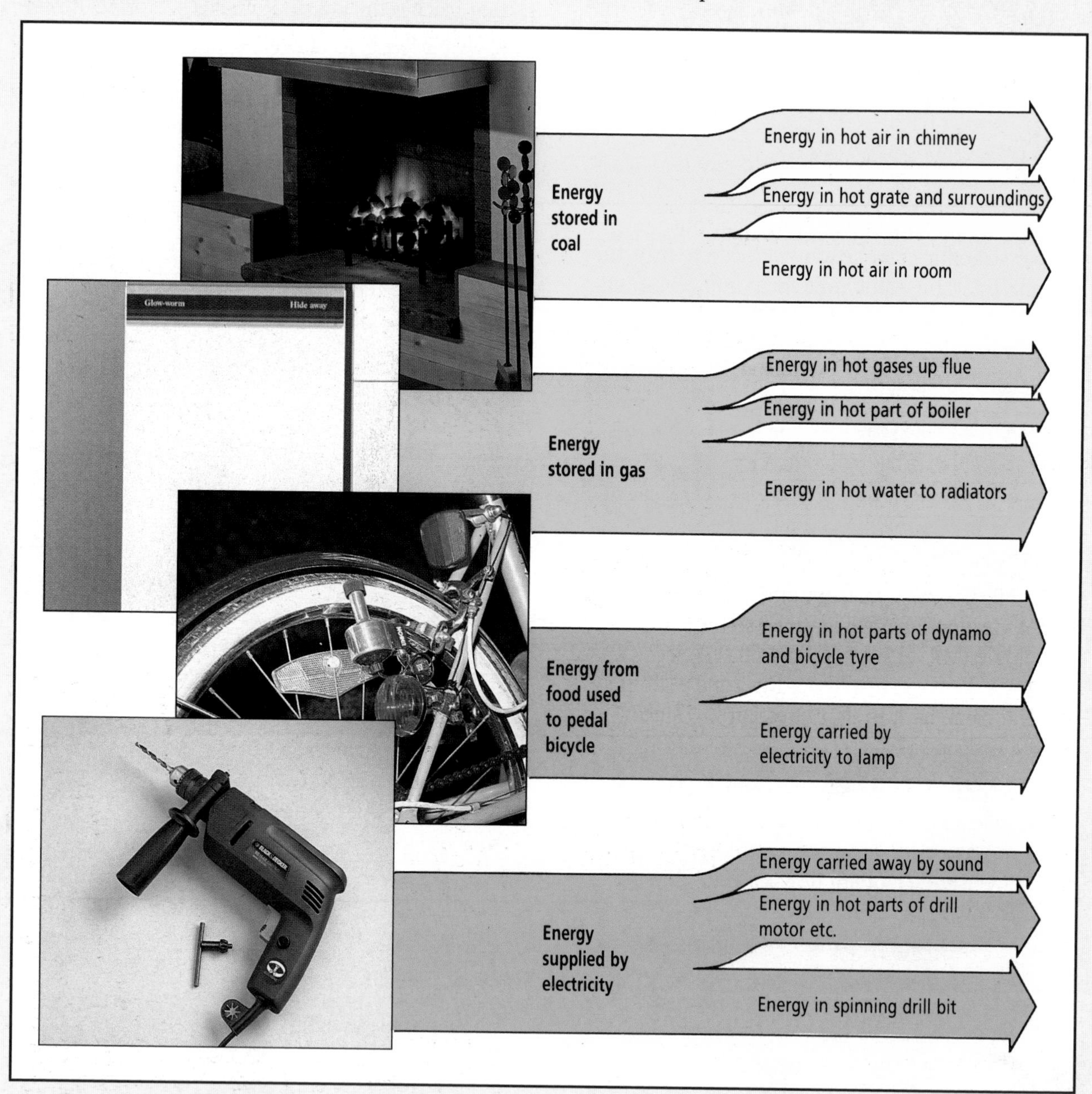

4. Energy patterns

Energy never disappears. And energy never suddenly appears from nowhere! A fuel is needed to make something happen, but the energy always appears somewhere at the end. Look at the energy arrow diagrams above. There is always energy on the output side as well as the input side. In fact, there is always *the same amount* of energy at the end as there was at the beginning – energy is **conserved**.

But the energy has been changed. It is now much more 'spread out'. Notice how the arrow diagrams always branch, with energy ending up in more and more places. You'll notice that in all the arrow diagrams some heating always happens, even when we don't really want it to.

5. Energy spreading

Not only do most events produce some unwanted heating, but the energy of hot things also goes on spreading! It is hard to keep energy concentrated in one place.

How does the energy in hot bodies spread? There are three main ways:

Conduction: The only way energy can spread through solids is by conduction. Metals are good conductors, but wood and plastics are poor conductors (good insulators). Water and air are also very poor conductors.

The copper bottom of the saucepan is a good conductor, so energy spreads easily from the hot cooker ring to the pan. The handle is made of plastic – a good insulating material – so that energy does *not* spread along the handle, making it too hot to hold.

Down is the best filling for high performance sleeping bags. Down traps lots of tiny pockets of air, which is a very poor conductor. So the energy cannot spread from your body to the surroundings and you stay warm.

Convection: When water is heated, it expands a little and gets lighter (less dense). So the hot water rises to the top, carrying its energy with it. Hot air does exactly the same. This method of energy spreading is called convection. Unlike conduction, the hot substance actually moves.

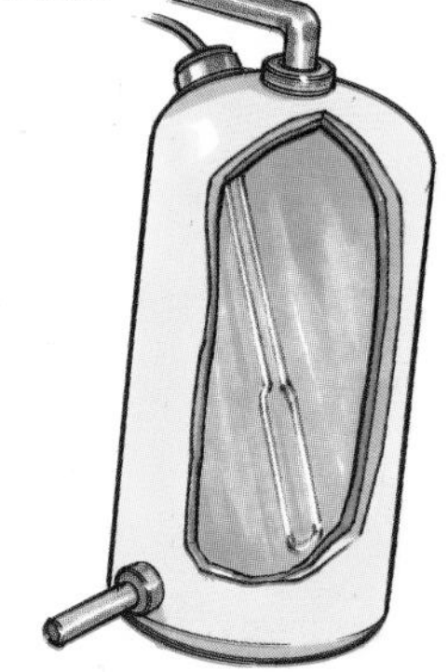

The immersion heater is placed near the bottom of the tank. It heats the water beside it, which then rises to the top of the tank. The pipe going to the hot taps leaves the top of the tank. Then the tank is refilled by cold water which goes in at the bottom.

Radiation: Energy spreads from the Sun to Earth through 150 million kilometres of empty space. It cannot spread by conduction or convection because there is nothing in between! So there must be a third method of energy transfer – radiation. Radiation travels very rapidly and in straight lines, just like light.

If you stand near a bonfire, you feel the intense heat on your face. You can shield it with your hand. The energy is spreading from the hot fire to your face by radiation travelling rapidly in a straight line.

6. Using energy efficiently

(a) We say that a process is *efficient* if most of the energy goes where we want it to go.

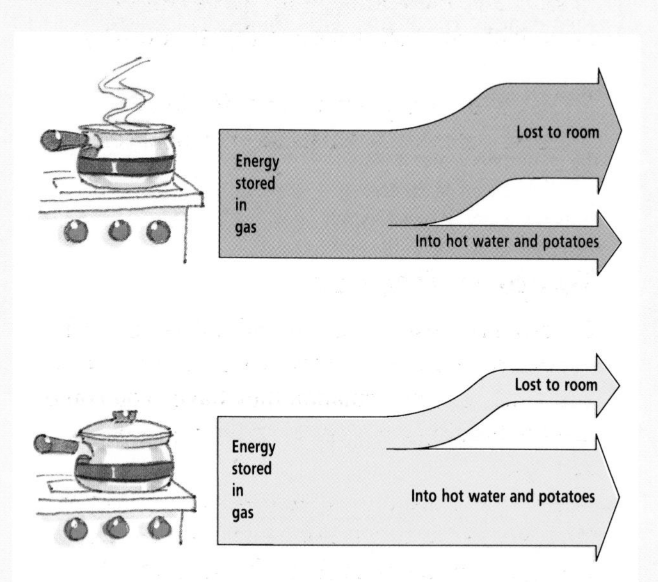

If we put a lid on the saucepan while we boil potatoes, it is more efficient. More of the energy from the fuel goes into the water and less is lost to the surroundings.

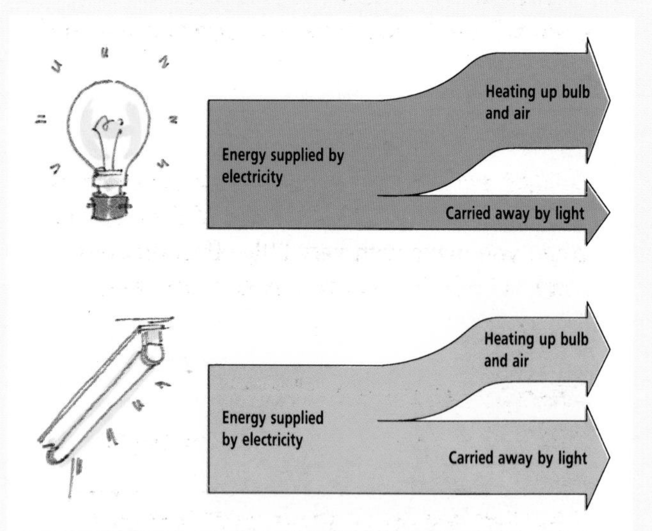

A filament light bulb is not as efficient as a fluorescent tube. Most of the energy which is supplied to the filament lamp just causes unwanted heating.

Efficiency is usually measured as a percentage: how much of the input energy goes where you want it to go?

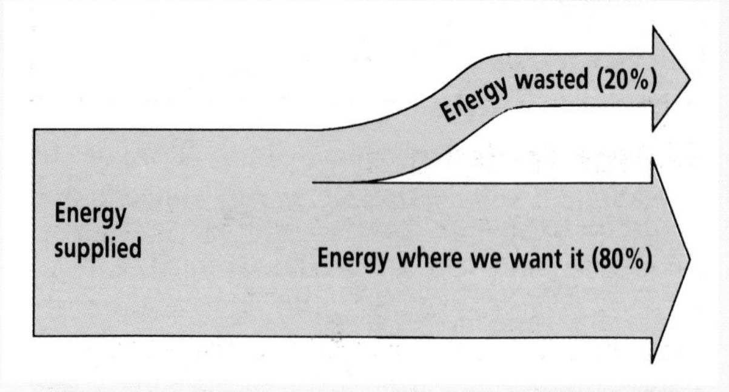

The process shown in the diagram above is 80% efficient.

We can now also explain why the three different methods of boiling water for a pot of tea (*Thinking about 2* on page 8) used different amounts of energy to do the same job. The water has to gain the same amount of energy each time. The electric kettle is the most efficient at transferring energy to water. The gas burners lose more energy to the surroundings.

(b) Insulating your house is a way of using the energy stored in the fuel efficiently.

A house made with good insulating materials will cost less to keep warm than one built with poor insulators.

The **U-value** of a material tells you how much energy will escape every second through each square metre if the temperature inside is 1°C warmer than outside. The bigger the area and the bigger the temperature difference, the more energy escapes every second!

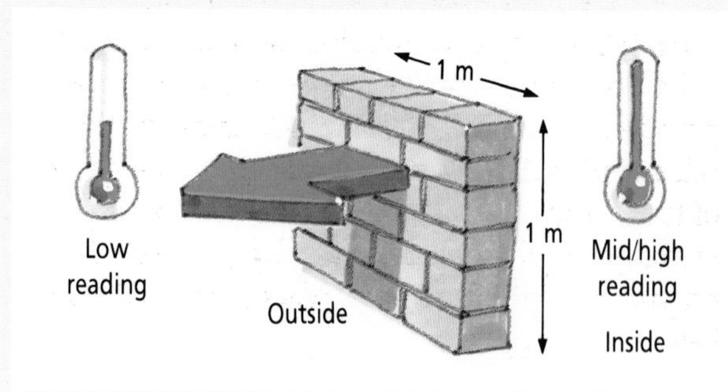

Material	U-value (in W per m^2 per °C)
Cavity wall	0·75
Insulated cavity wall	0·5
Uninsulated loft	2·0
Loft with 15 cm insulation	0·25
Single glazed windows	5·0
Double glazed windows	3·0

Things to do

Energy Matters

Things to try out

1 An energy-saving bulb (an SL bulb) is really a fluorescent tube coiled up inside a bulb.
It produces the same amount of light as a normal filament bulb – but uses less electrical energy.

The manufacturers claim that an 18 W SL lamp gives the same amount of light as a normal 75 W filament lamp – the same amount of light for only one quarter of the energy. Design two experiments to test these two claims and try them out.

Hint: One way of comparing the amount of light – the brightness – of two lamps is to use a solar-powered calculator. If you shine a lamp on the calculator's solar cell and then begin to lay sheets of tissue paper over the cell, it eventually stops working and the calculator display goes off. The brighter the light, the more layers of paper it takes before this happens.

2 (a) Watch the wheel turning inside the electricity meter in your house. Count the number of turns in 5 minutes. Work out how many turns this is equal to in an hour.

(b) The number of turns per kilowatt-hour will be written on the meter. Work out how many kilowatt-hours of electrical energy you are using in one hour.

(c) Now go round the house and make a list of all the appliances which are switched on and are using electricity from the meter. Find their rating in kilowatts and add them up. Does this figure agree with your calculation in (b)? If not, why?

3 Design and carry out an investigation to compare different materials for making sleeping bags.

Things to find out

4 Design a questionnaire to find out which fuels people in your class use for heating their homes, and what sorts of insulation they have. You could use the headings on page 2 for the types of insulation.

Use your questionnaire to conduct a survey in the class. Write a report on your findings, with tables and diagrams to show the main results.

5 Use the Yellow Pages of your local telephone directory to find the number of firms supplying different types of home insulation. Summarize your findings in a table.

What sort of insulation seems to be the most popular? Is it the most effective form of insulation?

Points to discuss

6 Make a list of the ways you could save energy **(a)** at home; **(b)** at school. Which of these savings could you make with very little effect on your comfort? Which could you make most easily?

Things to write about

7 This letter appeared in a local newspaper:

> **Another 'Save it' energy campaign. What a waste of time. Have the campaigners not heard about the conservation of energy? Energy is always conserved – you cannot waste it. There is no point in using energy more carefully.**

Write another letter, replying to the comments made in this one.

Questions to answer

8 Draw and label energy arrow diagrams (like the ones on page 10) to show the energy changes in each one of the following:

(a) a gas cooker

(b) a vacuum cleaner

(c) a candle

(d) a cassette-player

(e) a solar-powered calculator

(f) using the brakes on a bicycle

9 Look at the energy arrow diagram on page 10. Assuming it is drawn to scale, estimate the efficiency of each appliance.

10 'Who left the light on all weekend? That's two days that's been wasting electricity!'

'Sorry Dad! Anyway it's not as bad as you leaving the immersion heater on the other night.'

'That was only six hours – not nearly such a waste.'

Use the information on page 9 to settle the argument.

11 **(a)** The end wall of a house (see diagram) is 2 metres high and 5 metres long and has a U-value of 1·5 W per m^2 per °C. How much energy is lost per second through this wall when the temperature difference between the inside and the outside is

(i) 1 °C?

(ii) 10 °C?

(b) How much energy would be lost if it was an insulated cavity wall with a U-value of 0·5 W per m^2 per °C?

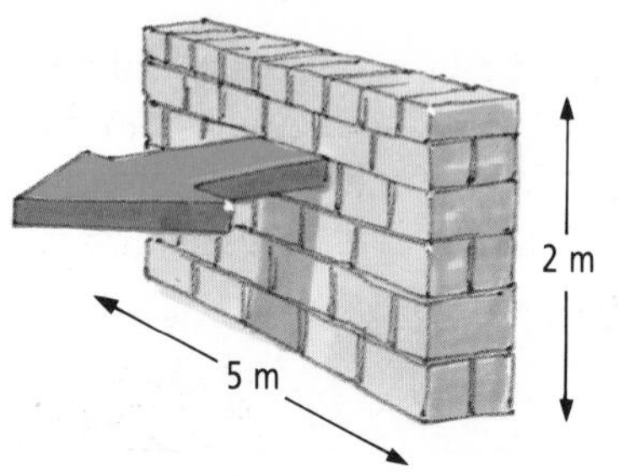

12 Make a table of the features of a vacuum flask and the type of heat loss each one prevents. Why is the stopper made of plastic?

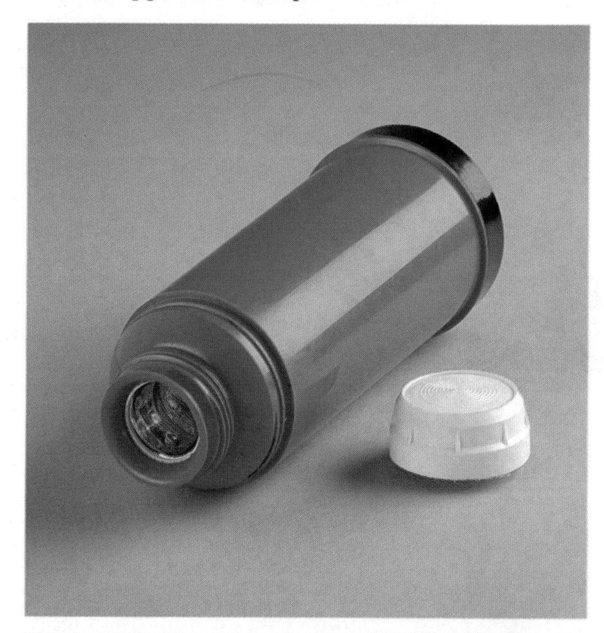

The vacuum flask keeps things hot by making it hard for energy to spread by any of the three methods: conduction, convection and radiation. The flask itself is a double-walled glass bottle, with a vacuum between the walls. And the inside surfaces of the glass bottle are silvered. The stopper is a potential weak link – but it is made of plastic.

Introducing

KEEPING HEALTHY

Your body is a very complex structure. To keep healthy you need to maintain it in good working condition. Keeping healthy also means knowing about what could go wrong, so that you can avoid problems.

1. You would know if your heart, stomach or bones had a problem – you would be able to feel something was wrong. Can you always tell if something is wrong with your body?
2. Does being healthy just mean not having a particular illness? Are you as healthy as a top-class athlete? What can you do to try and stay healthy?
3. If an athlete catches a virus it might make her feel a bit run down, but not too bad. Do you think this would affect her performance? Why?

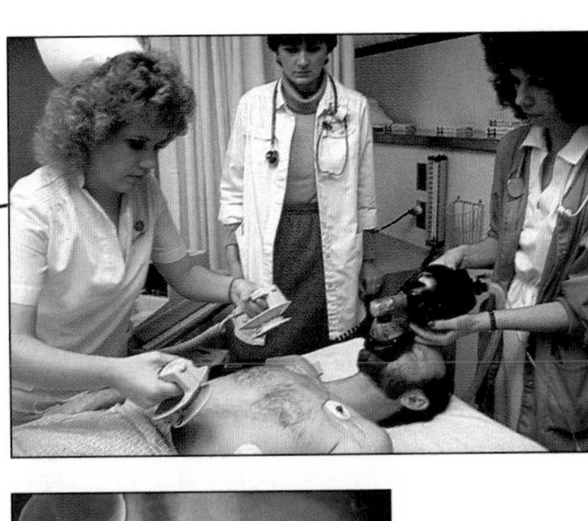

Your heart pumps blood around your body. The blood carries oxygen and food. If your heart goes wrong the cells in your body don't get the oxygen and food they need – you get seriously ill or may even die. This medical team is giving emergency treatment to a patient whose heart has stopped beating (he has had a cardiac arrest).

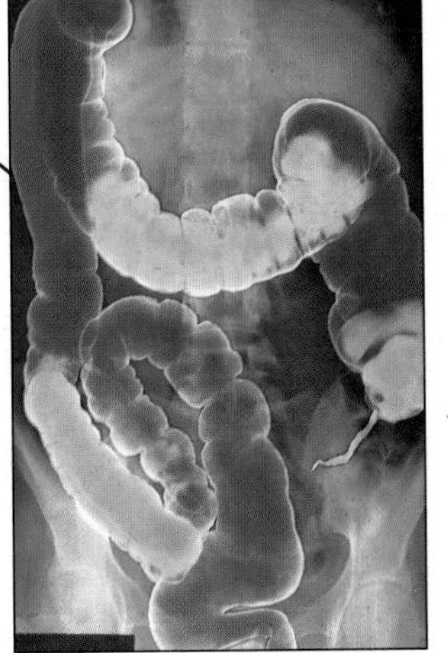

Your digestive system breaks down food so it can be absorbed into your blood. This is an X-ray of the large intestine taken after the patient swallowed a barium meal. The barium shows up on the X-ray.

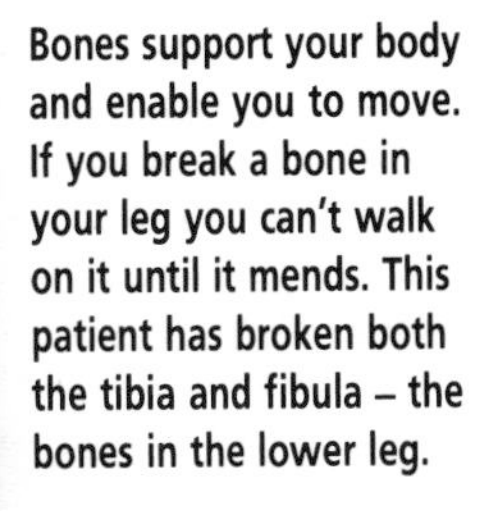

Bones support your body and enable you to move. If you break a bone in your leg you can't walk on it until it mends. This patient has broken both the tibia and fibula – the bones in the lower leg.

IN THIS CHAPTER YOU WILL FIND OUT

- what makes people ill
- what microbes are, and how the body deals with them
- what vaccination and immunity are
- how organs that are not working properly can be replaced
- how enzymes and drugs work.

Looking at

Influenza

'Flu away?

Have you had 'flu recently? Your symptoms would be a high temperature, a sore throat and a runny nose. Did you have to take any time off school?

During the winter, up to half a class can be away at one time due to a 'flu epidemic. Although most people are back to normal after a couple of days, 'flu can be very dangerous. In 1918 a worldwide 'flu epidemic caused the deaths of 21 640 000 people, one of the world's worst disasters.

In November 1989 there was a 'flu epidemic at Goodsalt school, as you can see from this register for class 4A.

NAME OF PUPIL		Week commencing Monday					1	
1	2	M	T	W	Th	F	Action taken on absence (initials and date)	Result of Enquiry
Surname	First Name(s)							
ADAMS	IAN	/ \	/ \	/ \	/ \	/ \		
BLOGGS	JANE	/ \	0 0	0 0	/ \	/ \		
CHOUDLEY	MEENA	/ \	0 0	0 0	0 0	/ \		
DAVIDSON	WINSTON	/ \	/ \	/ \	0 0	0 0		
EARBY	ERROL	/ \	/ \	/ \	0 0	0 0		
HARRISON	LUKE	/ \	/ \	0 0	0 0	/ \		
IYOTA	GEORGINA	/ \	/ \	/ \	/ \	/ \		
HO	SHUI LAI	/ \	0 0	0 0	0 0	/ \		
JOHNSON	LUCINDA	0 0	0 0	/ \	/ \	/ \		
KELLY	MARY	/ \	0 0	0 0	/ \	/ \		
KELLY	SARAH	/ \	/ \	0 0	0 0	/ \		
MAN	CHUNG FAI	/ \	/ \	0 0	0 0	/ \		
MATTHEWS	ROWENA	/ \	0 0	0 0	0 0	0 0		
NORMAN	NIGEL	/ \	/ \	/ \	0 0	0 0		
OWENS	WAYNE	/ \	/ \	/ \	/ \	/ \		
PATEL	NISHA	/ \	0 0	0 0	0 0	/ \		
PATEL	SUJATA	/ \	/ \	0 0	0 0	/ \		
PETERS	TOM	/ \	/ \	/ \	0 0	0 0		
RICHARDS	MARK	/ \	/ \	0 0	0 0	/ \		

1 Imagine you are Ms Anderson, the class teacher. Write a memo to the head teacher about the disruption to pupils' education that the 'flu epidemic has caused, and suggesting that something should be done to prevent it happening again.

Preventing 'flu

2 Now imagine that you are the deputy head who has been asked to reply to Ms Anderson's memo. The information you have gathered is on the rest of this page and the next one. Read it and answer the questions, then write your reply. (*In Brief* 6 and 7, pages 22–3, and *Thinking about* 2 and 3, pages 24–5, will also help.)

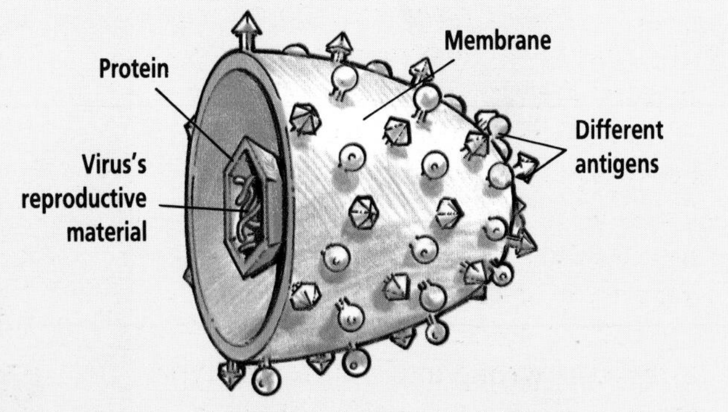

A model of a virus from the family Orthomyxoviridae ('flu virus). The spikes represent chemicals called *antigens*.

Influenza ('flu): a viral disease. It is very difficult to control because it is caused by not just one virus, but a whole family. To make matters more difficult, the viruses keep changing slightly.

As you can see from the picture, the virus is covered in spiky projections. These represent antigens. The body's immune system detects these and produces antibodies to attack the virus.

'Flu viruses are different from most other viruses because the antigens change slightly from one generation to the next. When the antigens have changed enough, your antibodies are no longer able to attack the virus. So as the 'flu viruses change, you lose your 'flu immunity. You become immune to most illnesses after having them once, but you may be re-infected with 'flu many times during your life.

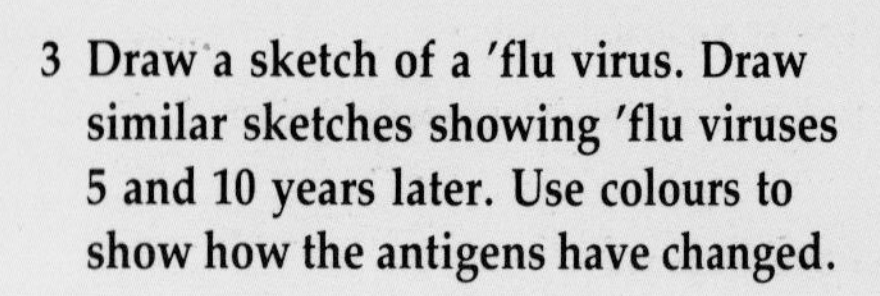

3 Draw a sketch of a 'flu virus. Draw similar sketches showing 'flu viruses 5 and 10 years later. Use colours to show how the antigens have changed.

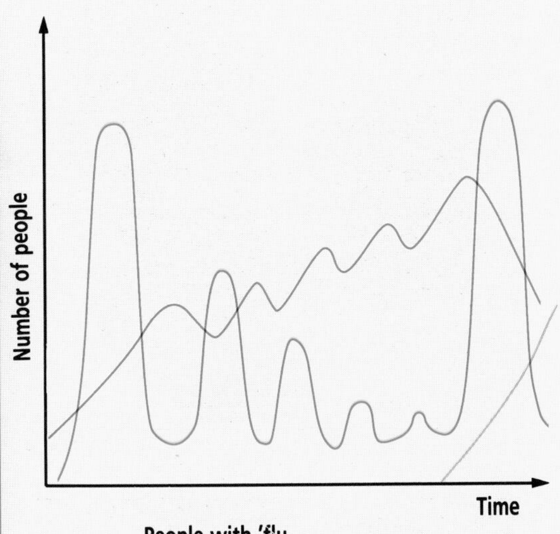

People with 'flu

Average number of people with 'flu antibodies

Average number of people with immunity to new strain of virus

'Flu immunity – the facts

The graph shows how outbreaks of 'flu are linked to changes in the virus. The red line shows the number of people suffering from 'flu. Gradually, after an epidemic, more and more people gain immunity to the disease. The number of people with 'flu immunity is shown by the blue line. Then a new strain of virus emerges, causing another epidemic. Immunity to this is shown in green. This immunity will build up following the previous pattern.

4 Predict the shape of the graph over the next ten years. Make a drawing of your predicted graph.

BEAT THE 'FLU – HAVE A JAB!

You can be vaccinated against 'flu – about 80% of people vaccinated do not catch it. The reason the vaccination is not always successful is that there are so many different kinds of 'flu viruses. Every February a group of doctors meets in Geneva to try to predict what 'flu viruses will be about in the following winter. Vaccines are then made up to protect against the predicted viruses – but if the forecast is wrong and a different type emerges, the vaccine will not work.

The mixture for the year 1990 protects against the following types of virus: A/Shanghai, A/Singapore, B/Yamagata. The first letter of the name is the type of 'flu virus; the second part is the place where that particular strain was first identified.

5 Imagine that a new type C virus has just been reported in your town. How will it be named?
What would happen if this virus began to infect people vaccinated with the 1990 mixture?

Looking at

Pellagra

What is pellagra?

Pellagra is a disease which dries and cracks the skin. It can be very painful. In severe cases the victim may die. In the 1900s there were so many cases of pellagra in the southern states of the USA that the government sent one of their top medical investigators, Dr Joseph Goldberger, to see if the disease could be controlled.

Dr Goldberger was already famous for helping to control two other dangerous diseases, typhus and yellow fever. Both of these were infectious diseases caused by microbes. Dr Goldberger had caught both diseases during his investigations – in fact he nearly died of yellow fever.

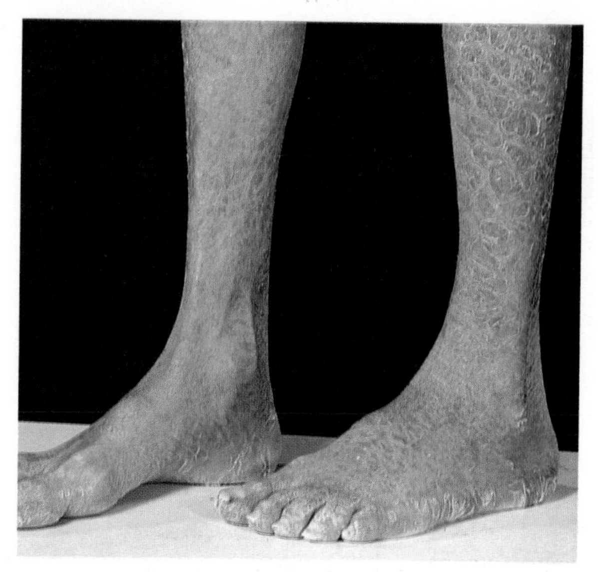

Pellagra is a very unpleasant disease. It was common in some parts of the USA about 100 years ago.

1 Imagine you are Dr Goldberger, on your way to the places where the disease has broken out. Make a list of the different types of diseases (*In Brief 1* on page 22 may help). Which type do you think pellagra might belong to?

2 What evidence would you need to show that pellagra belonged to this type of disease?

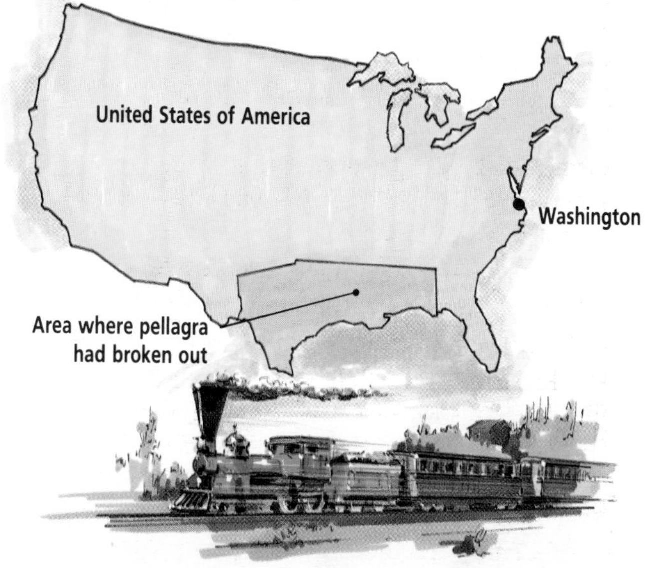

Was it caused by microbes?

Most doctors at the time thought that pellagra was caused by a microbe. They pointed out that many cases of pellagra were found in orphanages and mental hospitals, where large numbers of people were crowded together. These are ideal conditions for microbes to spread from person to person.

However, Dr Goldberger observed that only orphans and patients caught the disease. Doctors, nurses and other staff did not catch pellagra. This was unusual, and different from typhus and yellow fever, where all sorts of people caught the disease.

3 Write a page from Dr Goldberger's diary, setting down the reasons why you think pellagra is a different type of disease from typhus or yellow fever.

Crowded orphanages gave good conditions for microbes to spread.

Was it caused by poor diet?

Dr Goldberger wondered if the disease was caused by the poor diet that orphans and people in mental hospitals had. He arranged for many people suffering from pellagra to have fresh milk. Many of them quickly recovered. Was there something in milk that cured pellagra?

4 Write another extract from Dr Goldberger's diary, with the heading 'At last, a breakthrough'.

To get more evidence for his theory, Dr Goldberger arranged to conduct an experiment on a group of prisoners. They agreed to take part because they would be released from prison when the experiment was over. The diagram opposite shows what he did.

The prisoners were divided into two groups

Group A had the type of diet given to orphans and mental patients

Group B had the same diet as group A plus fresh milk

Half of group A caught pellagra

None of group B caught pellagra

Convincing evidence

Even after this experiment many doctors still thought that pellagra was infectious. To convince them, Goldberger performed an amazing experiment, using himself, his wife and a group of friends as part of the test.

If pellagra was infectious, then people who were in close contact with the disease should catch it. So, for nearly a year, Dr Goldberger and his group lived with pellagra sufferers. Not only were they in daily contact, they were also injected with blood samples from the diseased group. They also ate food to which they added faeces and urine from the diseased people!

And the result of this unusual experiment? Some of the group had stomach aches (is it any wonder!), but none of them caught pellagra.

5 Finish your extracts from Dr Goldberger's diary with an account of his experiments.

Why was milk important?

Milk contains nicotinic acid (also called niacin), a member of the vitamin B group. It is lack of nicotinic acid which causes pellagra. Meat and flour also contain nicotinic acid. New milling methods introduced around 1890 were so efficient they ground all the nicotinic acid out of flour – so more people were at risk of getting pellagra. Nicotinic acid is now added back into flour.

Joseph Goldberger

Looking at

Transplanting kidneys

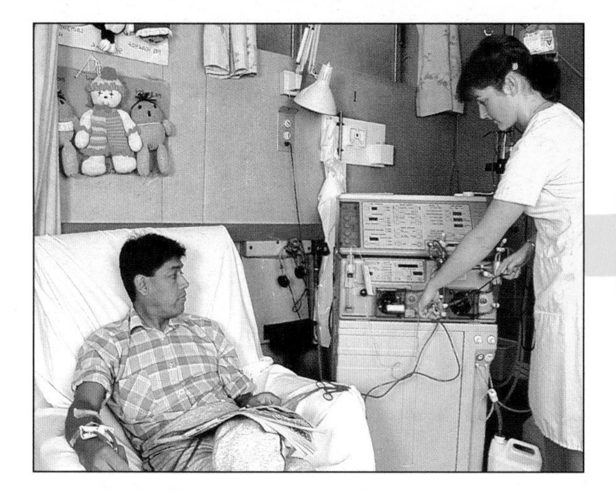

Abdul Khan is suffering from kidney disease. Both his kidneys have failed, so he has to have dialysis.

His sister, Ayesha, leads an active, healthy life.

Ayesha has decided to let Abdul have one of her two kidneys.

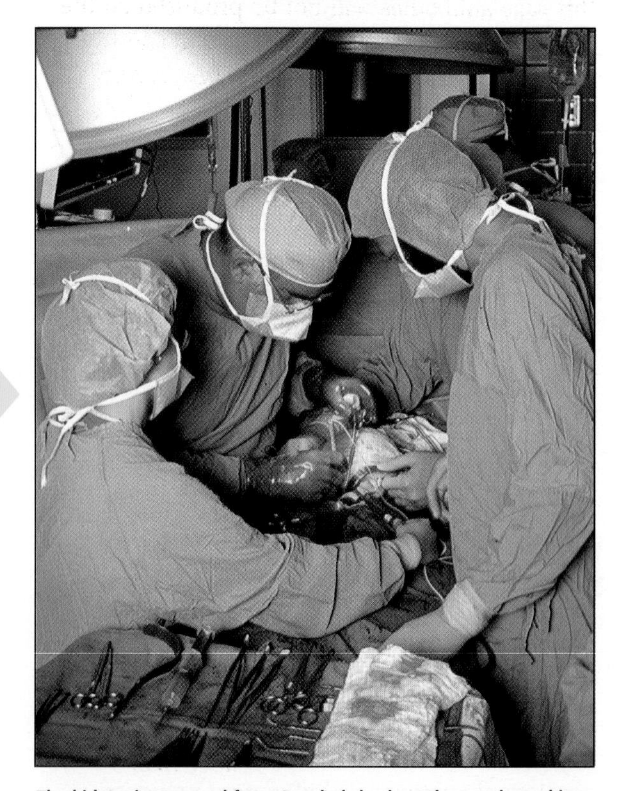

The kidney is removed from Ayesha's body and transplanted into Abdul's body.

Why did Ayesha have to give up one of her kidneys?

After all, sometimes people die and allow their kidneys to be used by other people (that's what kidney donor cards are for).

Transplants are not always successful. Sometimes the body **rejects** the new kidney. The immune system detects that the kidney is different from all the other tissues and starts to make antibodies. The antibodies surround and kill the new kidney, just as if it was a bacterium or a virus.

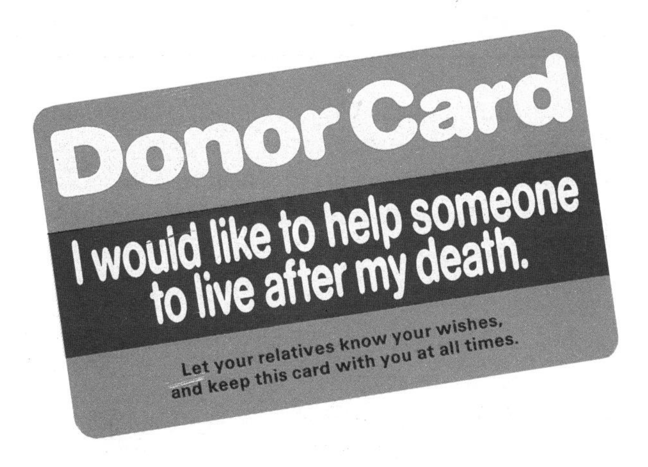

How can we stop the body rejecting a transplant?

One way is by using drugs. Some drugs stop the immune system working, so no antibodies are produced. The problem is that people have no defence if they catch a disease. The disease may kill them.

Another way of overcoming rejection is to use kidneys from people who are close relatives. The new kidney is similar to the other tissues already in the body, so there is a good chance that the immune system will not detect any difference. In this case antibodies will not be produced so the kidney will not be rejected. The graph shows how using kidneys from relatives improves the chances of success in transplant operations.

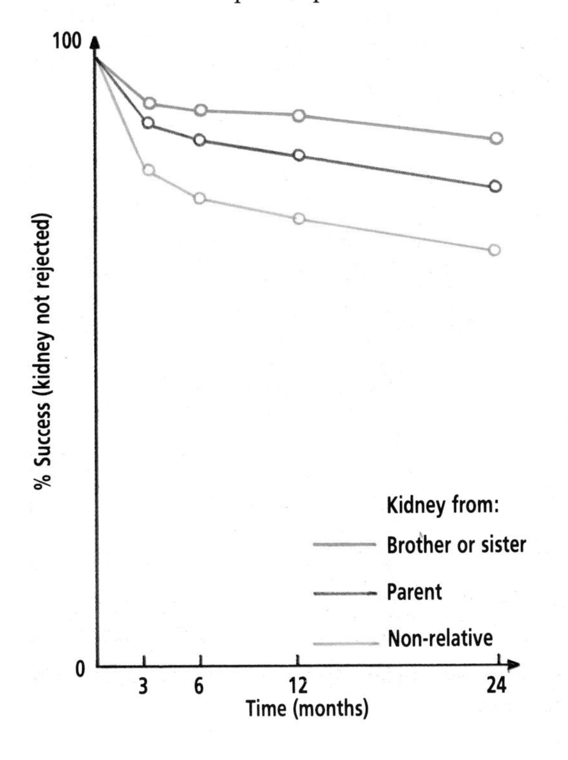

A year later . . .

1. What do you think antibodies react to in a donated kidney? Why are transplants of kidneys from close relatives likely to be more successful?
2. Abdul and Ayesha's parents were against the transplant to start with. They wanted Abdul to wait until a suitable kidney became available. Write a play to show how Abdul, Ayesha and their parents felt, and what changed their parents' minds.

In brief

Keeping healthy

1 Many complex chemical reactions go on inside your body. If any of these reactions goes wrong, you become ill. This may lead to unusual chemicals in your urine, so examining urine can help in the diagnosis of illness. Reactions can go wrong for various reasons:

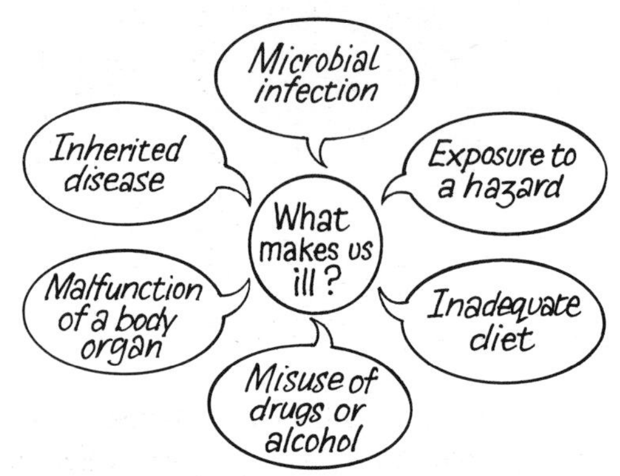

2 **Microbes** are tiny organisms that you need a microscope to see. **Bacteria** and **viruses** are microbes. There are many different types of microbes around us all the time, and your body makes a good environment for them to grow in!

Pathogenic microbes cause illness when they grow in the body. They can be passed on to other people and so infect them as well.

How microbes enter the body

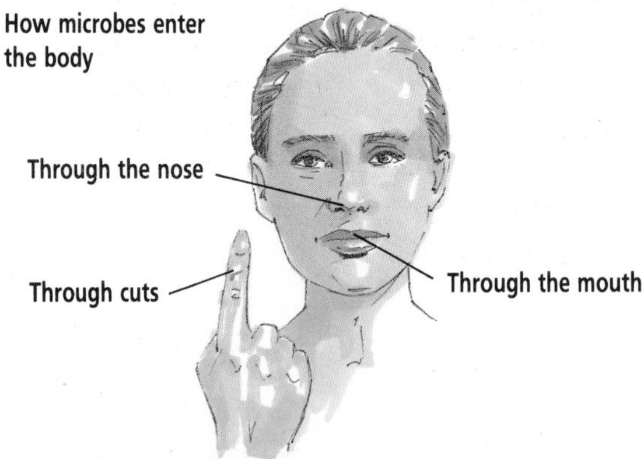

3 A microbial disease can be controlled by destroying the microbe or by eliminating the conditions it needs to grow. Microbes can be killed by:

- high temperatures
- acid or alkaline conditions
- removal of nutrients
- chemicals that poison them. These are called **germicides.**

4 **Disinfectants** contain concentrated germicides. They are only used on non-living material. **Antiseptics** are weak germicides that can be used on your skin. **Antibiotics** are germicides that are safe enough to be eaten or injected into the body.

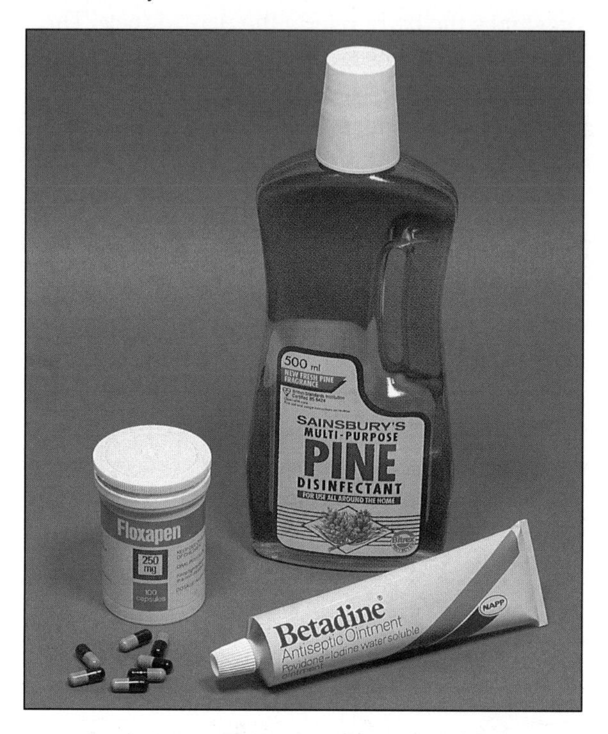

5 Your skin is a barrier to microbe invasion. When your skin gets cut, a blood clot forms to seal the wound and stop microbes entering.

6 When microbes do get into your body they are dealt with by the immune system. White blood cells produce antibodies which destroy the microbes. The next time you are infected with the same microbe, you already have antibodies to destroy it – you are **immune** to it.

White blood cells

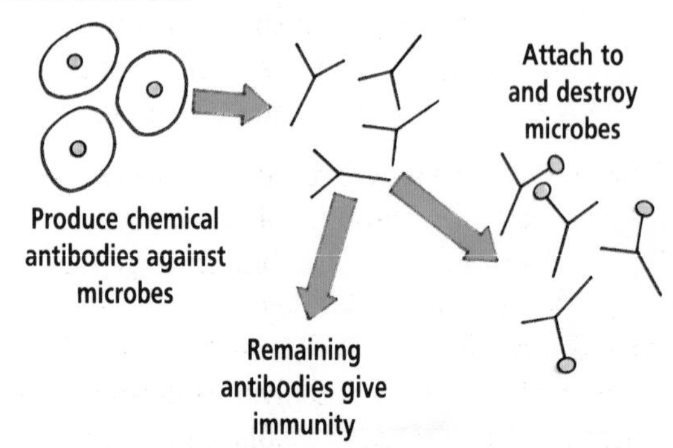

You become immune to a particular disease if you are vaccinated with a weakened form of the microbe. This stimulates the body to produce antibodies without making you ill. Immunization programmes have eliminated some major diseases.

7 You can also become immune by injection of antibodies.

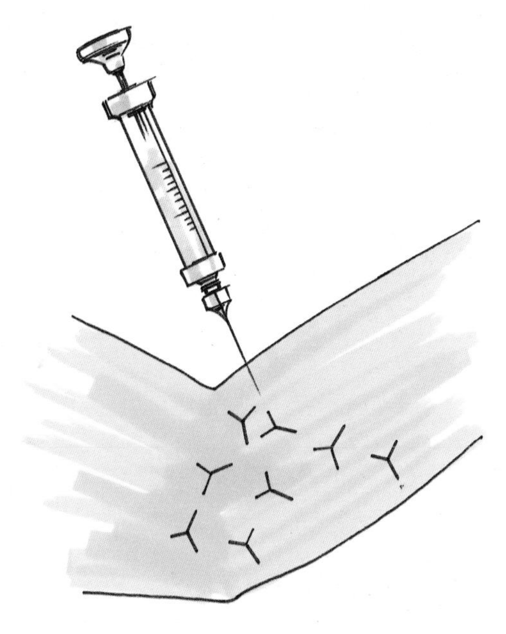

8 AIDS (acquired immune deficiency syndrome) results from infection by the human immunodeficiency virus (HIV). HIV attacks white blood cells and blocks the immune system by preventing antibody production. AIDS sufferers are unable to offer any natural resistance to microbial infection.

9 When body organs become diseased or do not function properly, they can often be replaced by healthy transplants. Kidneys, livers, lungs, hearts and corneas (from the eye) have all been transplanted. Although transplanted organs are carefully selected, they are often rejected by the body because of the immune response.

10 Your kidneys are high-pressure filter systems which control the amounts of water and dissolved substances in the blood. Although you are born with two kidneys, you can have a full and active life with only one. Kidney failure can be treated by regular dialysis (filtering) on a kidney machine, or by transplanting a healthy kidney from a donor.

11 Coronary heart disease happens when the blood vessels which supply the heart with oxygen get blocked. It can be treated by bypass surgery. An artery from the leg is transplanted to provide an alternative route to the heart.

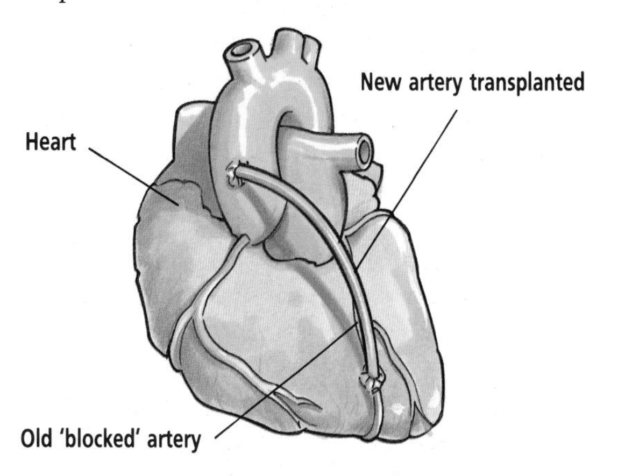

Valves in the heart keep the blood flowing in the right direction. Damaged heart valves may be replaced by artificial ones. Irregular heartbeats can be corrected with an electronic pacemaker.

12 **Enzymes** (biological catalysts) control the complex reactions that go on inside your body. To stay healthy, your body must keep the right conditions for enzymes to work in.

13 The chemical reactions which enzymes catalyse take place at an active site on the enzyme molecule. A substance which has a molecular structure that fits the active site can alter the action of an enzyme. Some drugs work by blocking the active sites of enzymes.

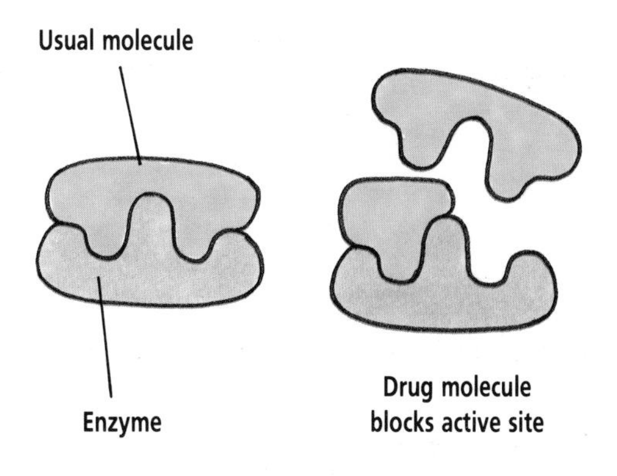

A drug that blocks the active site of an enzyme in a disease-carrying microbe will kill the microbe.

Thinking about

Keeping healthy

1. What happens inside your body?

Your body is made of lots of different chemicals. Like any other animal, you stay alive by taking in more chemicals. You take in water, oxygen and food (food is a mixture of chemicals). These react with other chemicals inside your body. Most of the reactions convert food chemicals into chemcials which are needed by your body and waste chemicals which are excreted from your body.

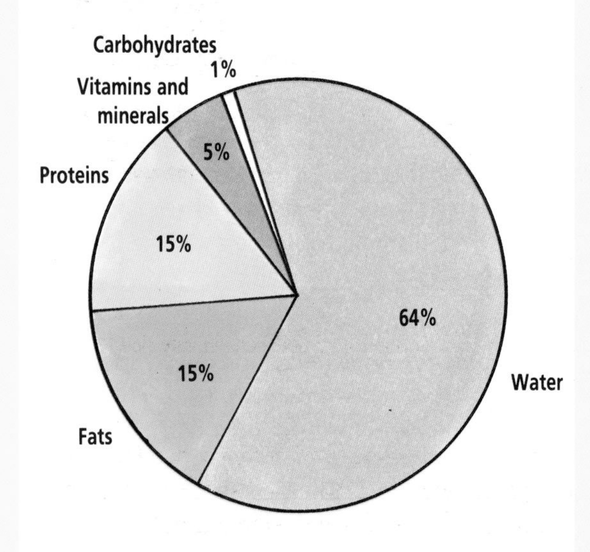

The chemicals which make up your body (percentage masses).

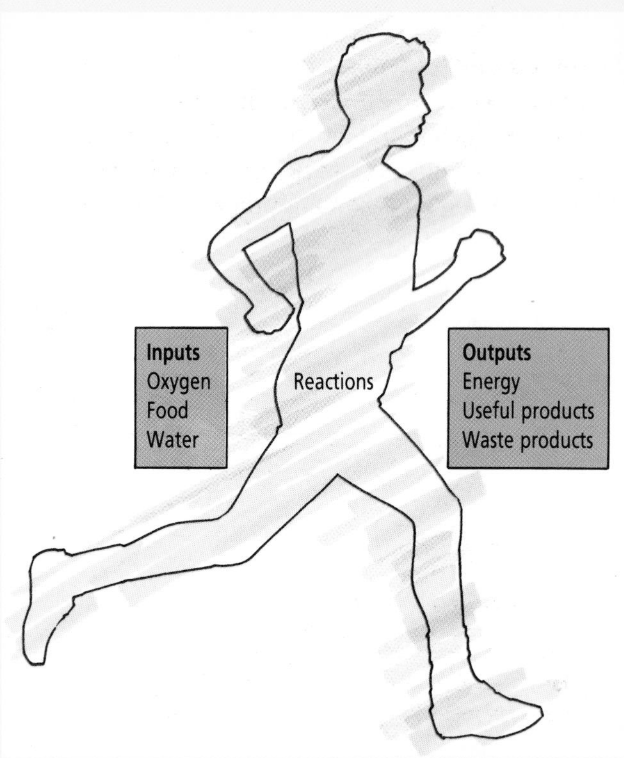

There are many different reactions taking place in your body. When these reactions are working properly, you are healthy. When they go wrong, you become ill.

These reactions are different from reactions you have seen in the laboratory. They are controlled so that they happen steadily, not all at once. The chemicals in the body that do this controlling are called **enzymes**. Enzymes catalyse reactions – they make them happen more easily. Some reactions would happen so slowly without enzymes that they wouldn't appear to react at all.

2. What are enzymes?

Enzymes are molecules that have a special shape. Each enzyme controls just one type of reaction.

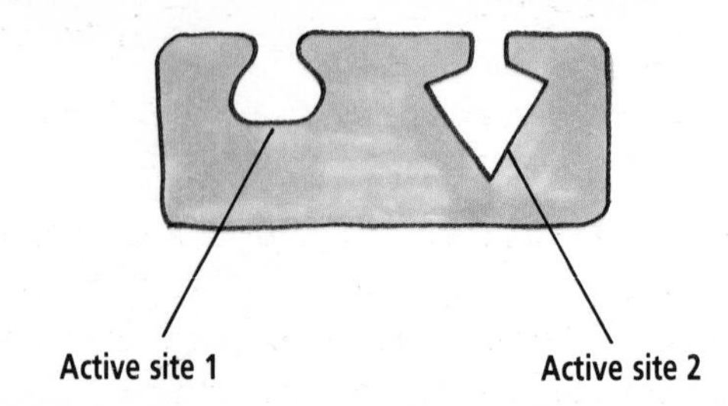

Enzymes have active sites – only one sort of molecule fits into each enzyme's active site.

Computer graphics image of the enzyme chymotrypsin. The active site is coloured.

How do enzymes work?

Enzymes are very sensitive to changes in conditions such as temperature and pH. These factors may change the shape of the molecule, so the enzyme no longer works properly. Microbes can also disrupt enzymes, by producing chemicals which fit into their active sites. When an enzyme is blocked your body chemistry is changed and you feel ill.

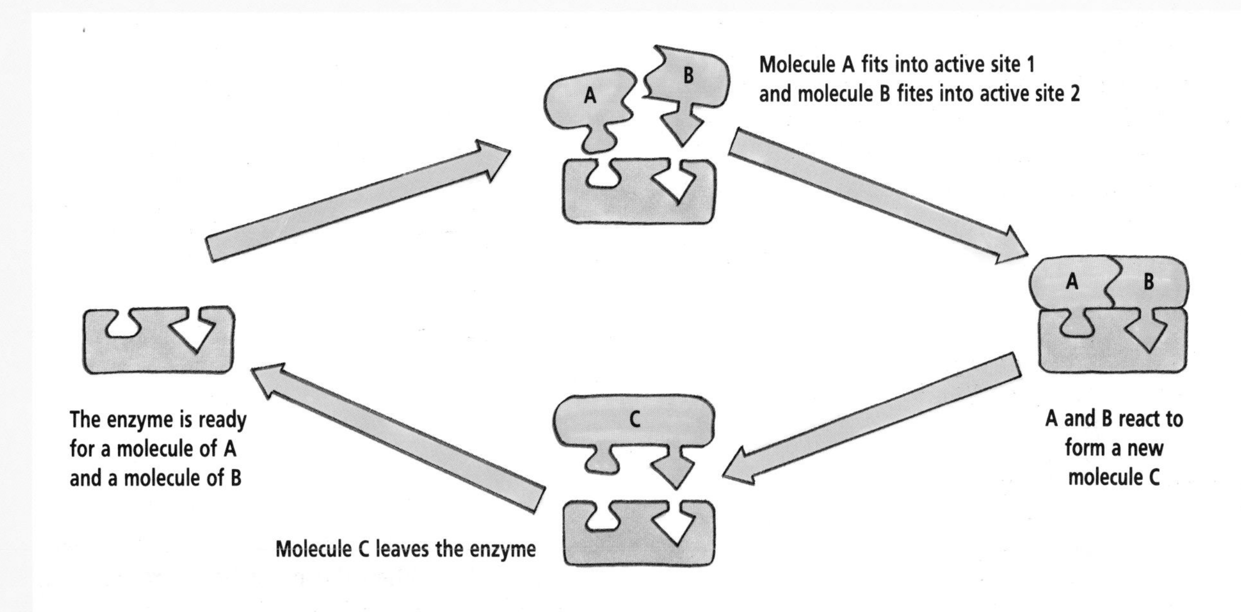

The enzyme holds A and B together so that C is produced easily. Without the enzyme the reaction would only happen very slowly.

3. How do drugs work?

Microbes are living organisms which have their own enzymes. Antibiotics can block the active site of an enzyme used by the microbe. The microbe's cell chemistry is changed so much that it dies, so the infection clears up. Because we don't have the same enzymes as microbes, the antibiotic doesn't harm our enzymes.

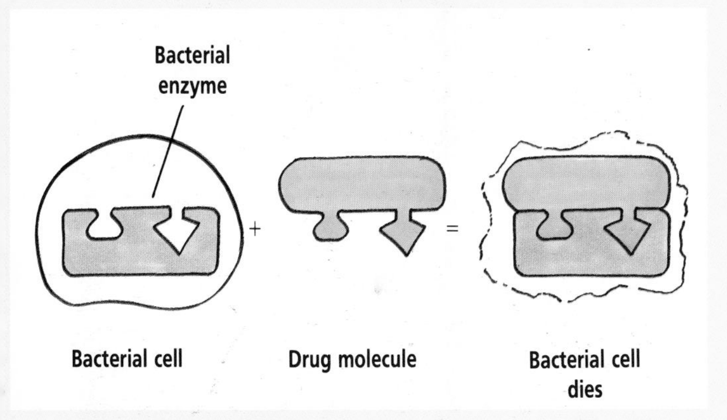

If a drug with molecules of a particular shape is effective against an illness, then a similar shaped molecule may also work. In fact, a similar shaped molecule may work even better. Chemists can design different drugs for a disease once they know the shape of one effective drug.

Enzymes are very complex molecules — this chemist is using a computer to design new drugs to block an enzyme.

However, developing a new drug is a long and costly process. A drug must be shown to be pure, safe and effective before it can be used. Even once it is allowed to be used, the drug must be monitored in case it causes any unexpected side effects. A new drug can take 20 years to develop, and the process may cost over £100 million.

4. *What are antibodies?*

Antibodies in your blood help you fight infection and so help keep you healthy. If you become infected with microbes, your body reacts by producing antibodies to them.

How do you produce antibodies?

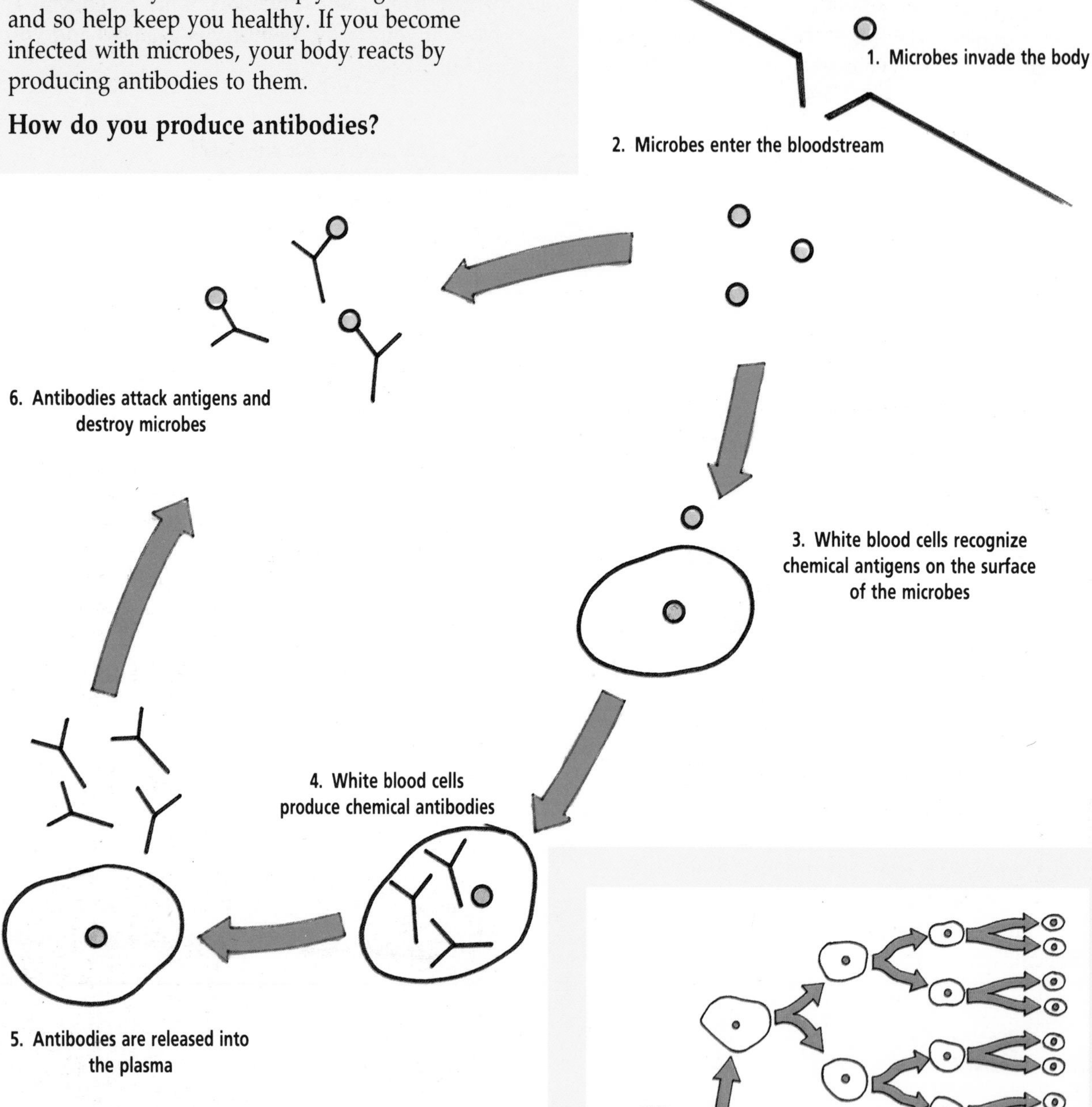

Antibodies are produced by white blood cells. When one white blood cell produces an antibody it enlarges and divides many times to form a large number of identical cells. Each cell is called a **clone**. All the clone cells can produce the same antibody, so there are lots of antibodies to fight the infection.

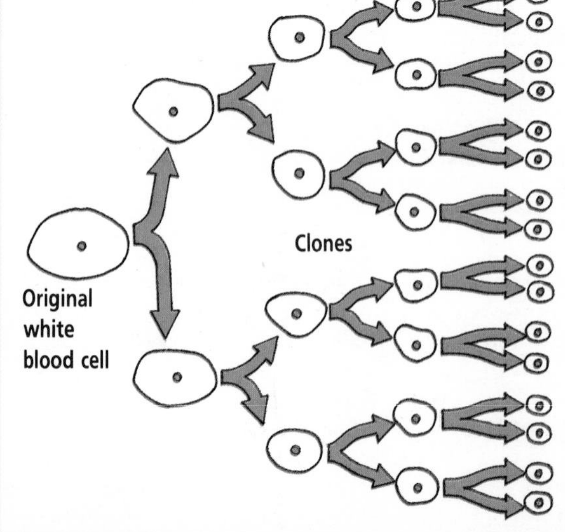

Clones are identical cells which are all formed from one original cell.

Once the infection has been controlled the antibodies are no longer needed and gradually break down. However a few of the white blood cell clones remain in the body. These act as 'memory' cells.

This is why you normally get a disease only once – the memory cells in your body give you lasting protection against the illness. This protection is called **immunity.**

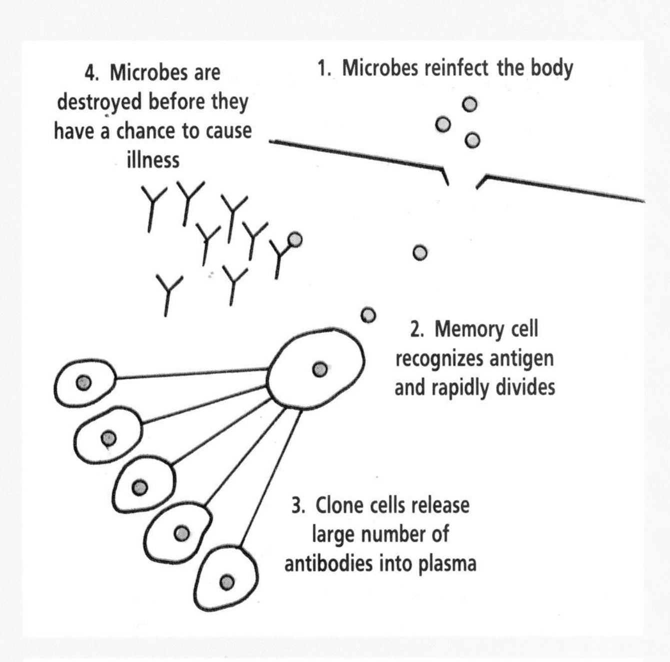

5. What is immunization?

Immunization is a process that makes your body 'remember' a disease so it can produce antibodies to it quickly. It can be done in two ways:

Active immunity

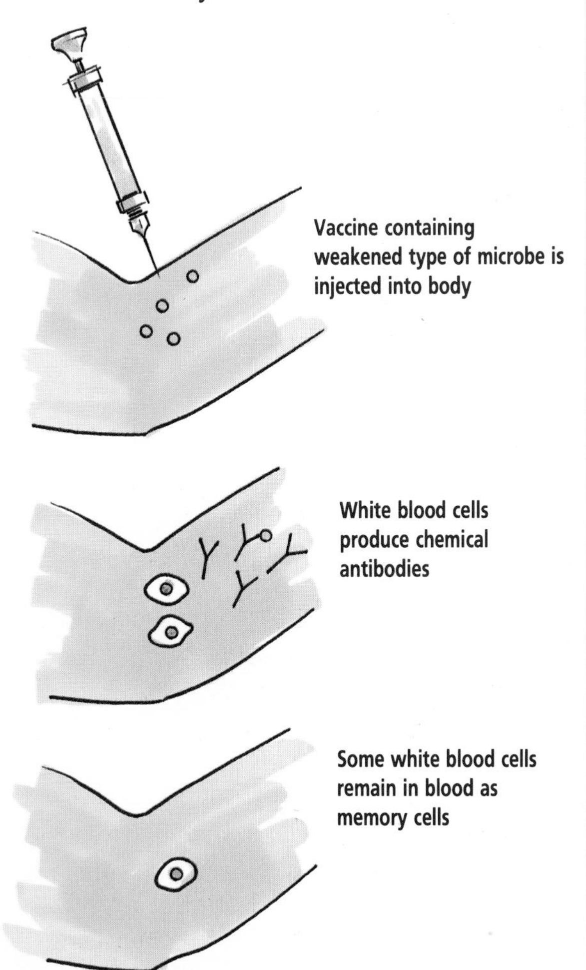

This is called active immunity because your body produces the antibodies for itself. Sometimes the vaccine consists of purified antigen, but you produce antibodies in the same way.

Passive immunity

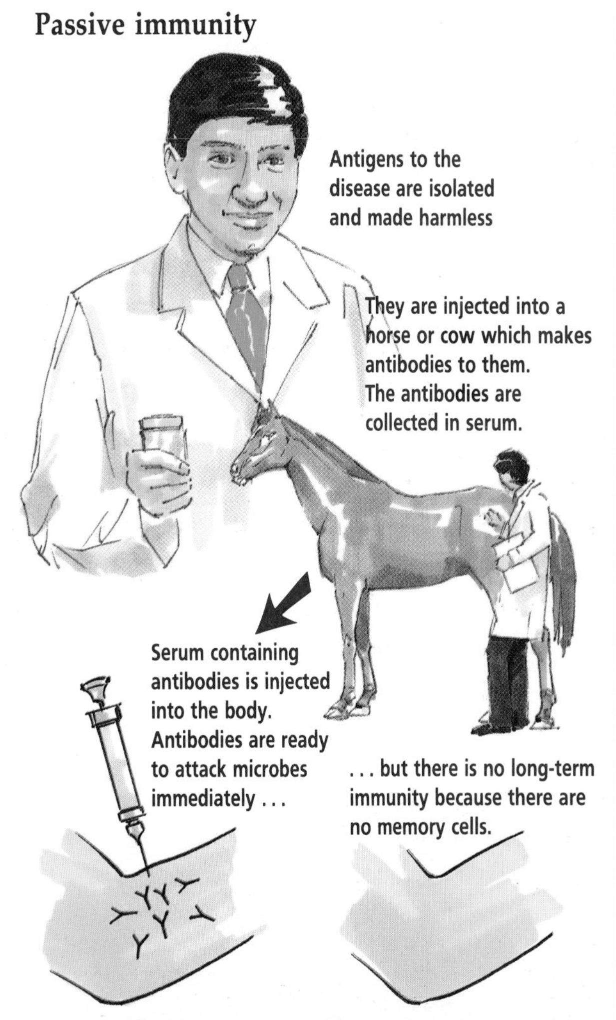

This is called passive immunity because the antibodies are injected into your body – it doesn't make them for itself.

6. How do kidneys work?

You have two kidneys which work around the clock to get rid of waste products from your blood. They filter about 1 litre of blood each minute.

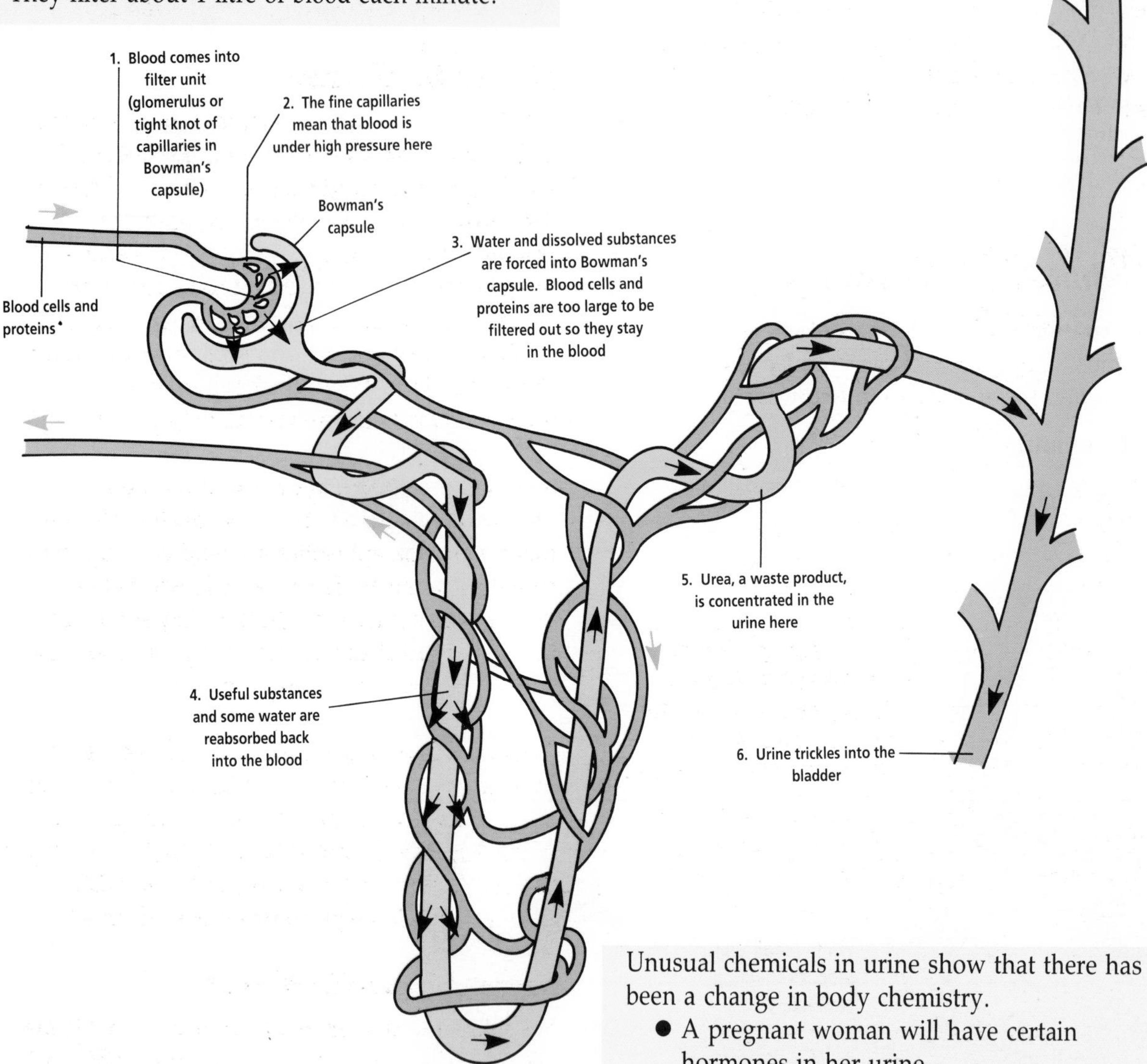

Healthy kidneys keep:

- the composition of your blood constant
- the water balance of your body constant.

If you drink a lot of water then your kidneys will not reabsorb all the water that passes through them back into the blood. You will produce a lot of dilute urine. If you lose water by sweating, your kidneys will reabsorb most of the water passing through them and so will produce a small amount of concentrated urine.

Unusual chemicals in urine show that there has been a change in body chemistry.

- A pregnant woman will have certain hormones in her urine.
- People suffering from diabetes may have glucose in their urine. People with diabetes lack the hormone insulin, which controls the level of glucose in the body. If the level rises too far glucose is excreted in the urine.
- People with a kidney disorder may have protein in their urine. Protein molecules are too large to be filtered through normal kidneys, but if the kidney is not functioning properly then proteins and other large molecules may be excreted in the urine.

Urine analysis is an important tool used in diagnosing illness.

Things to do

Keeping Healthy

Things to try out

1 Design a questionnaire to find out what common diseases people in your class have had, and what vaccinations they have had. Try out the questionnaire and present your results.

2

Use your knowledge of growing bacteria to design an experiment to test the truth of this statement.

Things to find out

3 Since the late 1940s cases of tuberculosis (TB) have been successfully treated by antibiotics. However, at that time, the disease was so common that there were national campaigns to prevent the disease spreading. Using other books, or by interviewing people who can remember, find out what happened during these campaigns.

Points to discuss

4 In 1985-6 there was a whooping cough epidemic and about 35 000 children caught the disease, which can sometimes lead to serious complications like pneumonia. A vaccine which prevents children catching whooping cough is available. A research study has shown that the vaccine can cause brain damage in one case out of every 100 000 vaccinations. Do you think young children should be given the vaccine? What other information would you need to help you decide?

5 Imagine that you have a sore throat, but otherwise you are healthy. Your doctor tells you that if you took antibiotics it would clear up the sore throat quickly. However, she will not give them to you because she believes they should only be taken when absolutely necessary. Do you think the doctor has done the right thing?

6 A drug has been developed that may be effective in fighting AIDS. However, it will take four to five years to show whether it is safe and whether it has any unexpected harmful effects. In that time many people may die of AIDS. Do you think AIDS sufferers should be given this untested drug?

Things to write about

7 Look at *Thinking About 5* on page 27. What is the difference between active and passive immunity? What are the advantages and disadvantages of the two types?

In the country of Tressla there has been an outbreak of the infectious disease typhoid. Both vaccine containing antibodies and serum containing antigens are available for typhoid. If you were a doctor, what advice would you give to the following people?

(a) a tourist going on holiday to Tressla next week

(b) a business person going to Tressla next month, who may make several more trips if the visit is successful.

Questions to answer

8

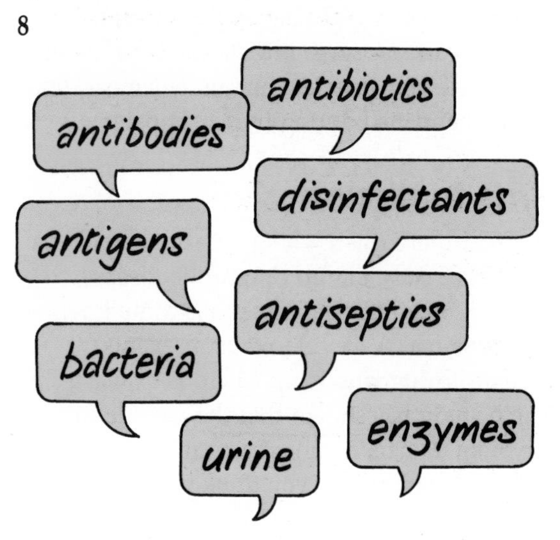

Copy and complete the following sentences. You can use these words more than once.

(a) __________ are a type of drug.

(b) __________ catalyse reactions in the body.

(c) __________ are living things.

(d) __________ are germicides that can be used on the skin.

(e) __________ can cause infections.

(f) __________ are germicides that should not be used on the skin.

(g) __________ are made by cells in the body to help the fight against infection by microbes.

(h) __________ is the name for the waste liquid that comes from the kidney.

(i) __________ are any substance foreign to the body.

9 The proportions of people dying from particular causes have changed a lot over the last 130 years. Table 1 gives the approximate percentages of deaths from four particular causes in 1850, 1910 and 1970.

Table 1

	Infectious diseases	Tuberculosis	Heart disease	Cancer	Other causes
1850	28	17	3	2	50
1910	15	11	9	8	57
1970	1	1	32	32	34

During this time the life expectancy (the averge number of years a person could expect to live) has also changed. Table 2 gives life expectancies, 1850 to 1970.

Table 2

Year of birth	Life expectancy (years)
1850	40
1900	46
1930	58
1970	70

(a) Display the data in each table as pie charts or graphs. Decide which way to present each set of information.

(b) Write a paragraph describing how the proportions of people dying from particular diseases have changed.

(c) Use the charts you have drawn to help you suggest an explanation for
- the increase in the proportion of people dying from heart disease
- the increase in the proportion of people dying from cancer.

10 A kidney machine uses dialysis to remove unwanted substances from a patient's blood. The blood is passed through a cellophane tube which is surrounded by moving fluid which contains dissolved salts. Use *Thinking About 6* on page 28 to help you to answer these questions.

(a) Explain which part of the kidney the cellophane tube corresponds to.

(b) Suggest why the fluid around the tube is moving.

(c) The tube allows water, salts, urea and ammonia to pass through it. Suggest why the fluid which surrounds the tube has salts dissolved in it before it is put into the machine.

(d) Suggest why the presence of protein in a patient's urine indicates a possible kidney disorder.

Introducing

TRANSPORTING CHEMICALS

What does the word 'chemical' suggest to you? Chemicals are not just bottles in the lab – everything is made up of chemicals! Sometimes we need to move chemicals from one place to another. The picture shows how four chemicals get from place to place.

1. List some advantages and disadvantages for each of these four methods of transporting chemicals.

Natural gas gets to the shore through a pipeline.

Petrol is transported from the refinery to the petrol station by road tanker.

This brewery still uses a horse and cart to take its beer to pubs.

Chemicals are moved from the factory where they are made to another factory by train. The second factory uses them to make products such as detergents.

IN THIS CHAPTER YOU WILL FIND OUT

- why chemicals are transported around the country and how chemicals are different from each other
- different ways of transporting chemicals and how to deal with chemical spillages
- the shorthand system to show the emergency services the hazards of the chemicals inside tankers
- the shorthand system of representing chemical substances which is used throughout the world
- how certain chemicals called elements can be classified.

Looking at

Why are Chemicals Transported?

Most things you eat or use have been manufactured – often from chemicals. The **chemical industry** takes raw materials – oil, gas, coal, minerals, air, water – and makes them into a wide range of chemicals. These chemicals are used to make anything from bread to compact discs.

1 How do you think the layer of rock salt which is under the ground in Cheshire got there in the first place? Write a paragraph showing your ideas.

Rock salt is an important raw material. This earth-moving vehicle is carrying rock salt in a salt mine.

The lorry takes the rock salt to the chemical works. Here it is converted into chlorine and sodium hydroxide.

2 Explain the advantages of siting the factory near the salt mine.

The rock salt is dissolved in water and electricity is passed through the solution. This converts it to sodium hydroxide and chlorine. The photograph shows a row of cells where this happens.

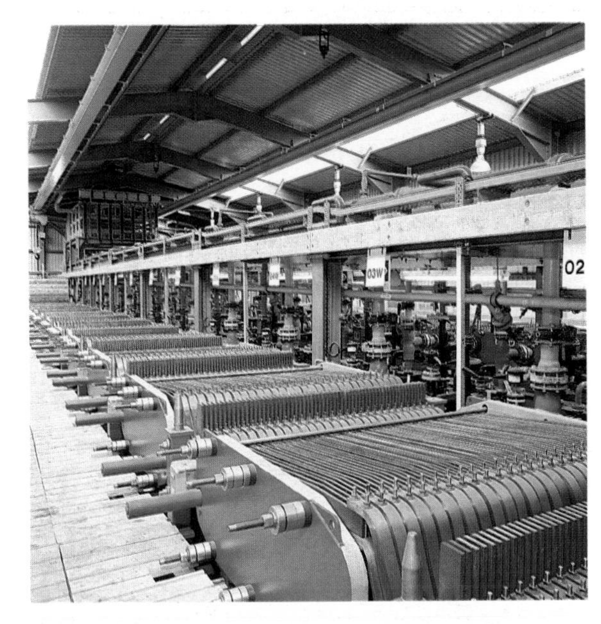

Chlorine and sodium hydroxide are very important chemicals – they are used to make many products which you use every day. But the factories which make these products are often a long way from Cheshire so the chlorine and sodium hydroxide have to be transported to them.

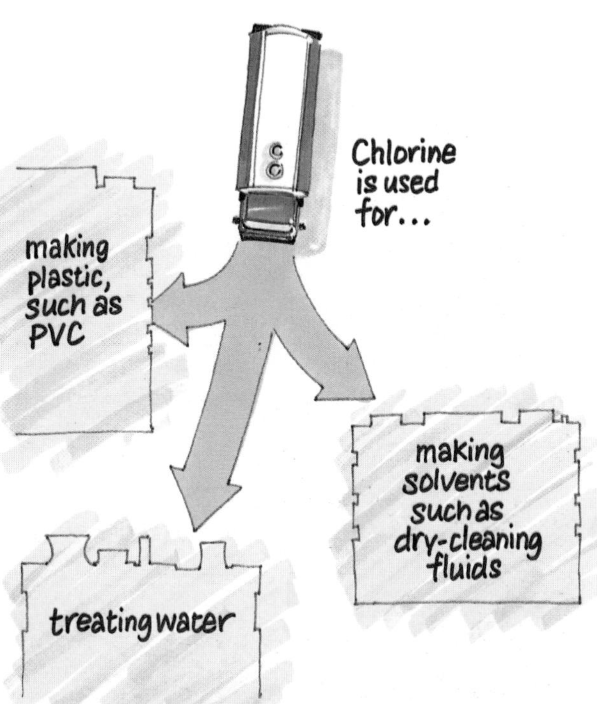

3 **Chlorine is a poisonous gas which changes easily into a liquid when you compress it. It is not flammable.**

Sodium hydroxide is a solid which absorbs water from the atmosphere. It is very soluble in water and its solution is very corrosive.

Bearing in mind these properties suggest the best way to transport each of these chemicals.

4 **Suggest why the paper factory which uses sodium hydroxide might not be near the factory which makes sodium hydroxide, and the factory which uses chlorine to make PVC might not be near the factory which makes chlorine.**

Looking at

How Can We Classify the Elements?

1789

It was the year of the French Revolution. A French nobleman called Antoine Lavoisier published a book in which he classified the *elements.* He did this by grouping elements with similar properties together. He might have made further progress with this but the revolutionaries chopped off his head with the guillotine in 1794!

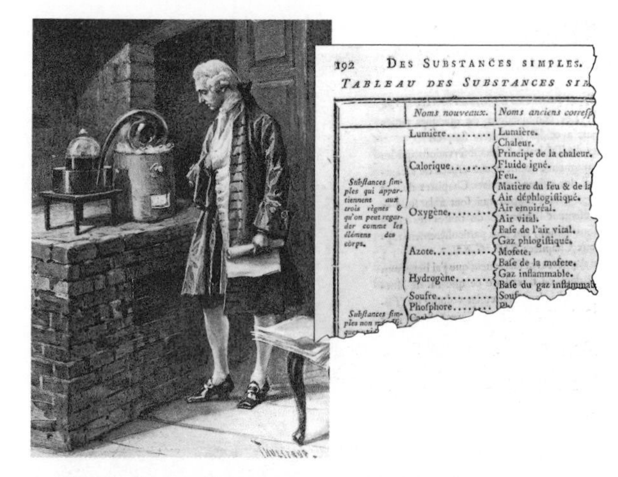

192 DES SUBSTANCES SIMPLES.

TABLEAU DES SUBSTANCES SI

	Noms nouveaux.	*Noms anciens corresp*
Subſtances ſimples qui appartiennent aux trois règnes & qu'on peut regarder comme les élémens des corps.	Lumière..........	Lumière.
	Calorique..........	Chaleur. Principe de la chaleur. Fluide igné. Feu. Matière du feu & de la
	Oxygène..........	Air déphlogiſtiqué. Air empiréal. Air vital. Baſe de l'air vital.
	Azote..........	Gaz phlogiſtiqué. Mofete. Baſe de la mofete.
	Hydrogène..........	Gaz inflammable. Baſe du gaz inflammab
Subſtances ſimples non	Soufre..........	Souf
	Phoſphore..........	

Lavoisier and one of his groups. As well as the elements oxygen and nitrogen (azote) it contained light and heat.

1863

The story continued when John Newlands, a British chemist, arranged all the known elements in the order of the increasing masses of their *atoms.* He noticed that when they were arranged like this there was a pattern in their properties. At a meeting of the Chemical Society in London he announced that:

> ... the eighth element starting from a given one is a kind of repetition of the first, like the eighth note of an octave in music.

The pattern became known as Newland's Octaves.

This pattern or generalization only worked for the first 16 elements. This fact, plus the way he linked the pattern to musical notes, led some of the other chemists at the meeting to ridicule his idea.

John Newlands

The repetition of similar properties at regular intervals is called **periodic variation.** Although other people did not accept Newlands' generalization it was the beginnings of the **periodic table of the elements** which scientists now accept and understand.

1 Give a reason *why* scientists wanted to classify the elements.

2 If John Newlands was right, which element would you expect to have properties similar to (a) lithium (b) beryllium?

1869

Six years later, Dimitri Mendeleev, a Russian, published another periodic table. The basic idea of his table was the same as Newlands in that he

- arranged the elements in order of the masses of their atoms
- and put elements with similar properties under each other in the same vertical column

but where an element did not seem to fit he left a space. He predicted that another element would be discovered with properties which would fit these spaces.

Mendeleev was using his generalization or pattern to **predict** the properties of unknown elements.

For example, he realized that arsenic (As) fitted better under phosphorus (P) than under silicon (Si), so he left a gap (?) under silicon. He predicted this undiscovered element would form an oxide which would be white and have a high melting point and in which one atom of the element would combine with two atoms of oxygen.

1884

Fifteen years later the missing element was discovered. It is called germanium. Germanium forms a white compound with oxygen which has a formula of GeO_2 and a melting point of 1389 °C. Mendeleev was right so chemists began to believe that the periodic table was useful and had some underlying **explanation.**

Mendeleev did not suggest an explanation for the periodic variation of properties. This happened much later, in this century, when **theories** about the structures of atoms were developed.

Dimitri Mendeleev was Professor of Chemistry at St. Petersburg University (now called Leningrad).

	Group 1	Group 2	Group 3	Group 4	Group 5	Group 6	Group 7	Group 8
Period 1	H							
Period 2	Li	Be	B	C	N	O	F	
Period 3	Na	Mg	Al	Si	P	S	Cl	
Period 4	K Cu	Ca Zn	? ?	Ti ?	V As	Cr Se	Mn Br	Fe Co Ni
Period 5	Rb Ag	Sr Cd	Y In	Zr Sn	Nb Te	Mo I	?	Ru Rh Pd

3 A magazine which is published at regular intervals, such as once a month or once a year, is called a periodical. Why do you think the table of elements is called a *periodic* table of elements?

4 Write a short newspaper article reporting on the Chemical Society meeting at which Newlands announced his 'Law of Octaves'.

5 Lavoisier, Newlands and Mendeleev all published their ideas. Why do you think scientists publish the results of their experiments, and the theories or generalizations based on their results?

6 Mendeleev's periodic table is a generalization or pattern. You can make predictions from it but it is not a theory. Use the ideas discussed on this page to suggest why this is so.

Looking at

Cargoes

Here is a famous poem about transporting chemicals (and other things).

Quinquireme of Ninevah from distant Ophir
Rowing home to haven in sunny Palestine
With a cargo of ivory,
And apes and peacocks,
Sandalwood, cedarwood, and sweet white wine.

Stately Spanish galleon coming from the Isthmus,
Dipping through the Tropics by the palm-green shores,
With a cargo of diamonds,
Emeralds, amethysts,
Topazes, and cinnamon, and gold moidores.

Dirty British coaster with a salt-caked smoke stack
Butting through the Channel in the mad March days,
With a cargo of Tyne coal,
Road-rail, pig-lead,
Firewood, iron-ware, and cheap tin trays.

(John Masefield)

1 **Which of the cargoes mentioned in the poem do you think are pure chemicals?**
2 **The poem was written over 50 years ago. Write a fourth verse about the sort of cargoes carried nowadays.**

In brief

Transporting Chemicals

1 The chemical industry is a large and important part of the manufacturing industry. Sometimes it converts raw materials into products which we buy, such as oil into petrol. But often it first converts the raw materials into chemicals (intermediates) which are then used to make products.

2 The factories used to make products are built at different places.

The choice of site is influenced by:

- source of raw materials
- road, rail and sea links
- available workforce
- environmental factors
- energy supply
- water supply.

3 The intermediate or bulk chemicals can be transported to other factories by:

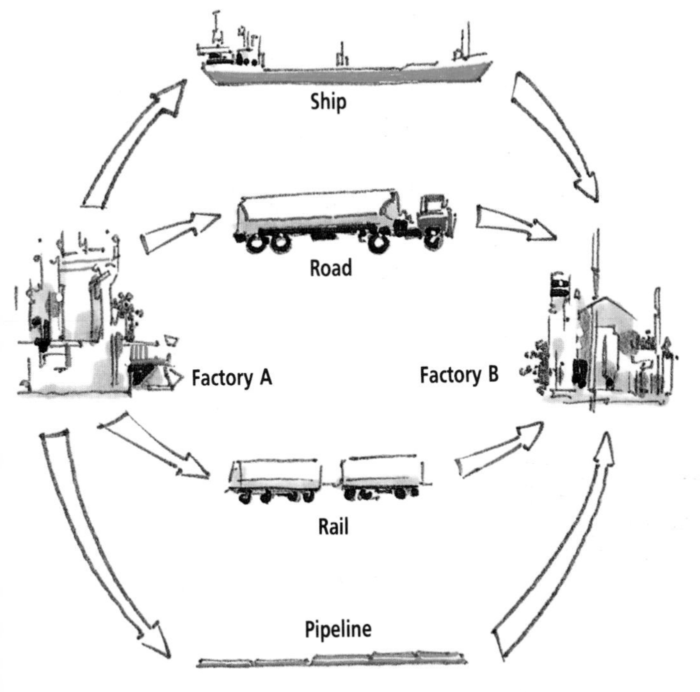

The method used depends on geographical, economic, social and environmental factors.

4 Different chemicals often have different properties. For example, some burn, some are corrosive and some are gases. When they are transported around the country their containers must be carefully labelled. The **Hazchem code** system of labelling tells the police and fire services dealing with an emergency what sort of chemicals are involved.

5 An **element** is a substance which cannot be broken down into anything simpler. The smallest particle of an element is called an **atom.** Atoms of one element are different from atoms of all other elements. Each element is represented by a **symbol**, which is shorthand for the element.

Compounds contain the atoms of more than one element joined together (not just mixed up). Each pure compound is represented by a formula which is shorthand for the compound.

For example,

$$CaCO_3$$

represents calcium carbonate. This tells you that it contains

calcium, carbon and oxygen

and that these elements are present in the ratio

1 : 1 : 3

The smallest particles of some compounds and elements are called **molecules**. For example, H_2O represents a molecule of water and H_2 a molecule of hydrogen.

6 Chemical reactions between elements or compounds can be represented by word equations. They list the starting substances (reactants) and the products. For example, the burning of natural gas (methane) can be represented by:

methane + oxygen → carbon dioxide + water

Alternatively, an equation which is made up of symbols and formulas can be used:

$$CH_4 + 2O_2 \rightarrow CO_2 + 2H_2O$$

This is called a **balanced** equation because it has the same number of atoms of each element on each side.

7 The **periodic table** is a way of arranging and displaying the elements which helps you remember the similarities and differences between them. It collects similar elements together in vertical columns called **groups**. The horizontal rows are called **periods**.

Thinking about

Transporting Chemicals

1. How should chemicals be transported?

If you ran a factory making chemicals you would sell them to other factories which turn them into useful products. How would you decide on the best way of getting the chemicals to the other factories? You might choose road tankers, trains, ships or pipelines. For small amounts which need to get there quickly you might even use a plane.

Which method do you think is cheapest? You want to make a profit, but the cheapest way might not be the best. If your factory is near a town or village the people won't want your lorries thundering by. There might not be a railway near – you could have one built, but this would be expensive and so you'd have to charge more for your chemicals.

You have to think about lots of questions – here are some for getting chemical X from factory A to factory B.

To predict the cost of a method of transport you need to consider both capital costs and running costs.

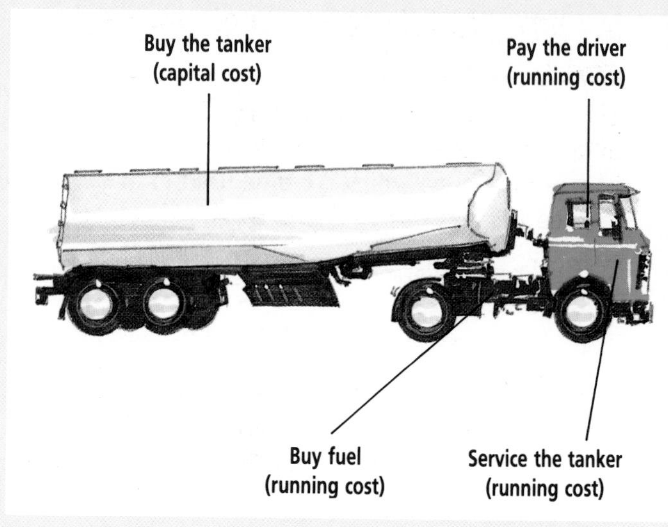

A pipeline to transport liquids or gases would be very expensive to build (capital cost) but cheap to maintain (running cost). It would be worth it if you had to transport a chemical from your factory to another for a number of years.

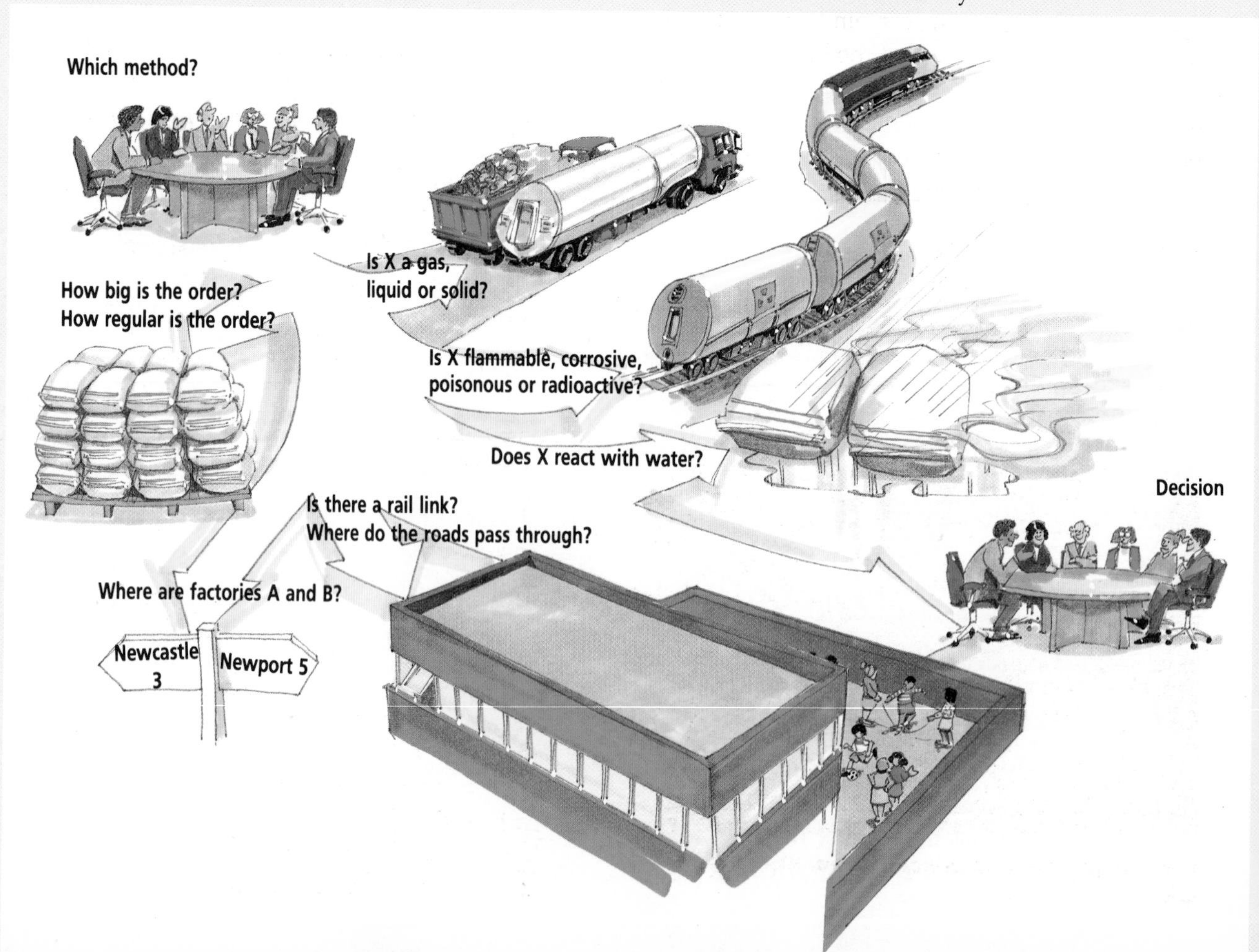

2. *Why do we need to know how chemicals differ?*

There are millions of known chemicals which are different in many ways. As a first step it is important to know about a few ways they can differ rather than remembering the properties of a particular chemical.

Chlorine is a pale green, poisonous gas which does not burn

Hydrogen is also a gas, but it is colourless, it is not poisonous and it forms an explosive mixture with air

Alcohol is a colourless, flammable liquid which mixes with water

Petrol is also a colourless, flammable liquid but it does not mix with water

Sodium carbonate is a white solid which dissolves in water to form a harmless solution

Sodium hydroxide is also a white solid which dissolves in water but it forms a very corrosive solution

If a lorry or train carrying a chemical has an accident, the police and fire services need to know about the properties of the chemical. So there is an agreed system for labelling containers and vehicles carrying chemicals. It is called the **Hazchem code**.

The emergency workers can look up the name of the chemical from the chemical code number. But the number and letters in the top box are more important – they tell them how to deal with a spillage.

The key shows what the number and letters mean. The 2 means that fog fire equipment can be used. The letter P means that the substance can react violently, the fire fighters should wear full protective clothing and the chemical can be diluted with water and washed down a drain.

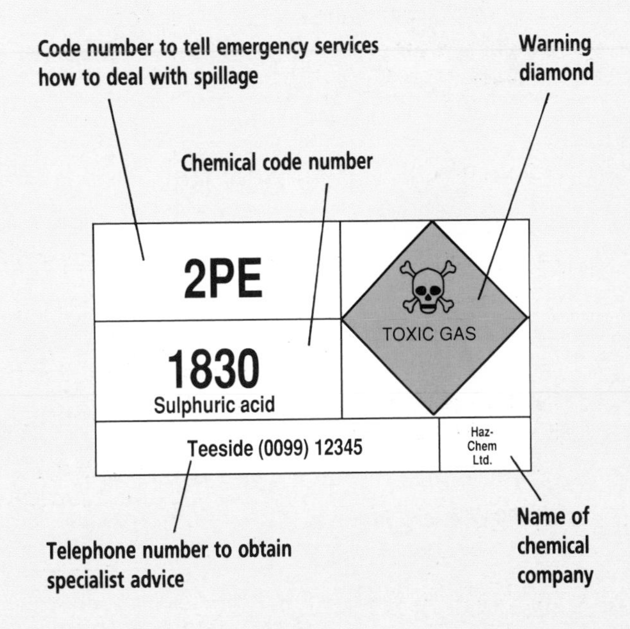

In an emergency the Hazchem code gives vital information quickly.

Hazchem Scale

P	V	FULL	DILUTE
R			
S	V	BA	
T			
W	V	FULL	CONTAIN
X			
Y	V	BA	
Z			
E	CONSIDER EVACUATION		

1 JETS
2 FOG
3 FOAM
4 DRY AGENT

3. What are elements, compounds, symbols and formulas?

People in the emergency services who do not usually know a lot about chemistry can understand the Hazchem code (see page 39).

Scientists all over the world also have a shorthand system. It is used to represent *all* pure chemical substances and gives a lot of information.

Every material object, living and non-living, in the whole universe is made of one or more **elements**.

There are about 100 elements. The arrangement of them shown below is called the **periodic table**. Each element is represented by a **symbol**.

Each element consists of tiny particles called **atoms**. The atoms of one element, for example copper, differ from the atoms of all other elements.

The atoms of different elements can combine to form substances which are called **compounds**. The smallest parts of compounds are called **molecules**.

Compounds are represented by **formulas**. The formula is made up of the symbols of the elements which are in the compound and small numbers which show in what ratio the atoms of the elements are present.

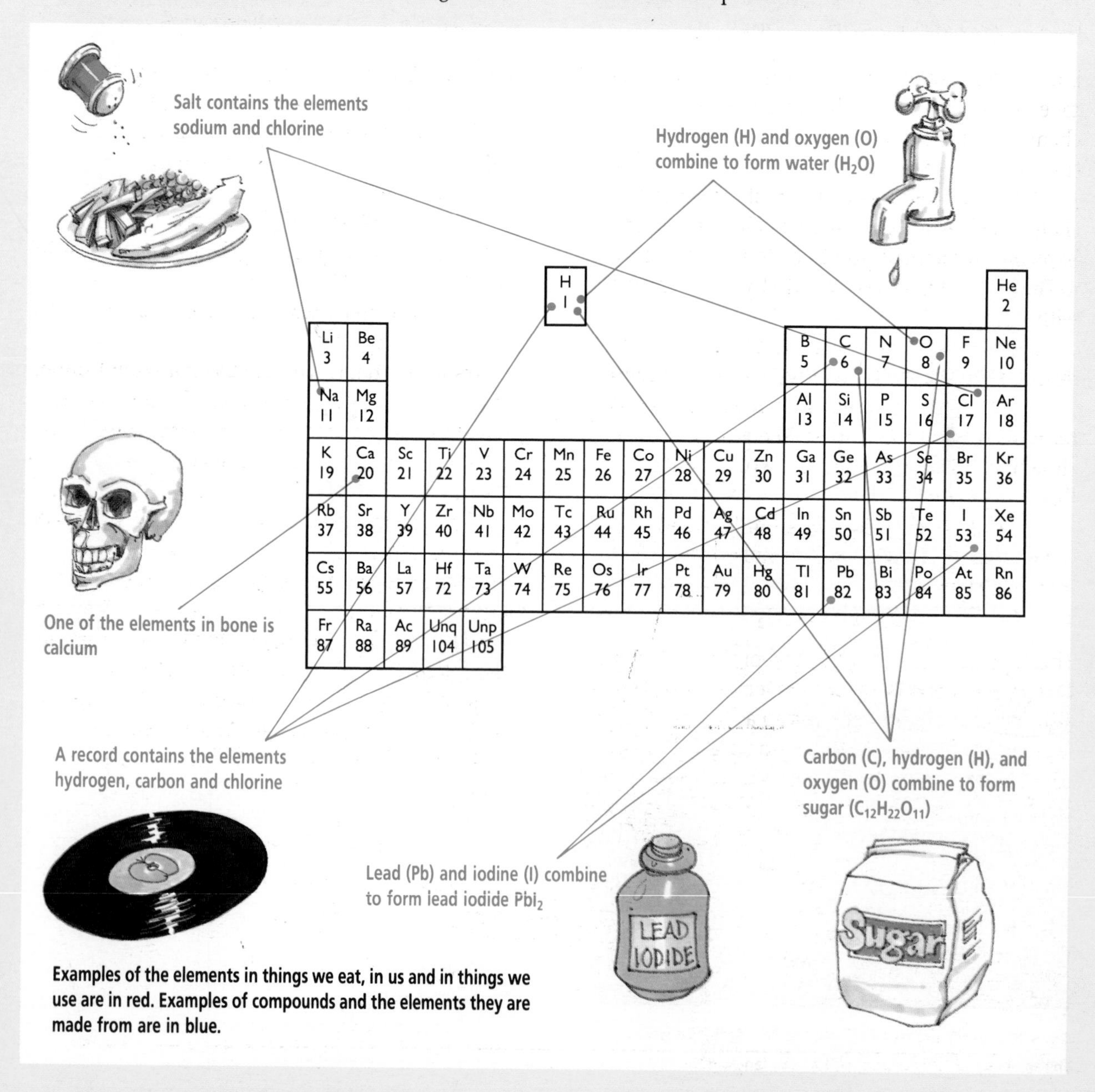

Examples of the elements in things we eat, in us and in things we use are in red. Examples of compounds and the elements they are made from are in blue.

4. How are symbols and formulas used to describe chemical reactions?

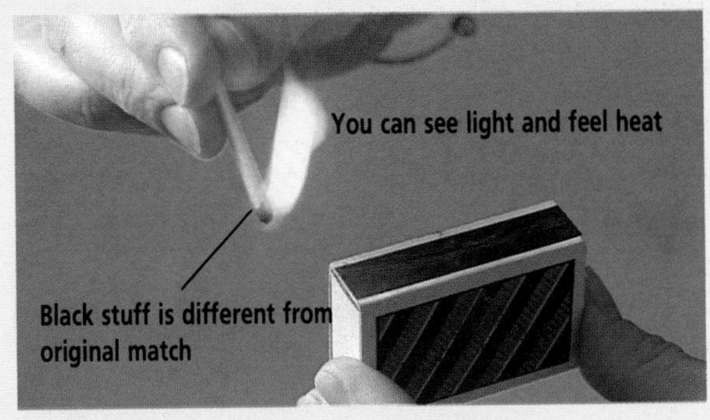

When a match burns new substances are formed.

When people make wine new substances are formed.

Both of the changes shown in the photographs are called **chemical changes** because the starting chemicals, the **reactants**, have changed to different chemicals, the **products**.

You cannot always be sure that a chemical change has taken place by just observing what happens. Sometimes you have to test what is left to show it is different from what you started off with.

When you work out what is happening you can describe the change by writing a **word equation**.

For example, if a piece of charcoal (carbon) is burnt in oxygen you can test the gas formed to show it is carbon dioxide. The word equation for this change is:

carbon + oxygen → carbon dioxide

Sometimes symbols and formulas are used instead of words:

$$C + O_2 \rightarrow CO_2$$

This equation tells you that carbon is an element. C represents one atom of the element. Oxygen is also an element. O represents an atom of it but the formula O_2 shows that these atoms go around in pairs. Carbon dioxide is a compound because it contains two elements combined together and the formula CO_2 shows that they are combined in the ratio of one carbon atom to two oxygen atoms.

Overall the equation tells you that one C combines with one O_2 to form one CO_2.

The equation:

$$2Mg + O_2 \rightarrow 2MgO$$

tells you that: 2 atoms of magnesium combine with one O_2 to form 2MgO.

When a chemical reaction takes place, no atoms are created and none are lost. They are just rearranged. That is why the 2 is placed in front of the Mg and in front of the MgO. The total number of atoms of each element on the left is equal to the total number on the right. The equation is **balanced**.

If you look in science books you will see lots of these equations. They are a quick way of describing chemical reactions and they provide more information than word equations. They are used by scientists throughout the world – they are the international language of scientists.

At this stage it is more important that you understand the information that the equation is telling you rather than be able to write equations yourself.

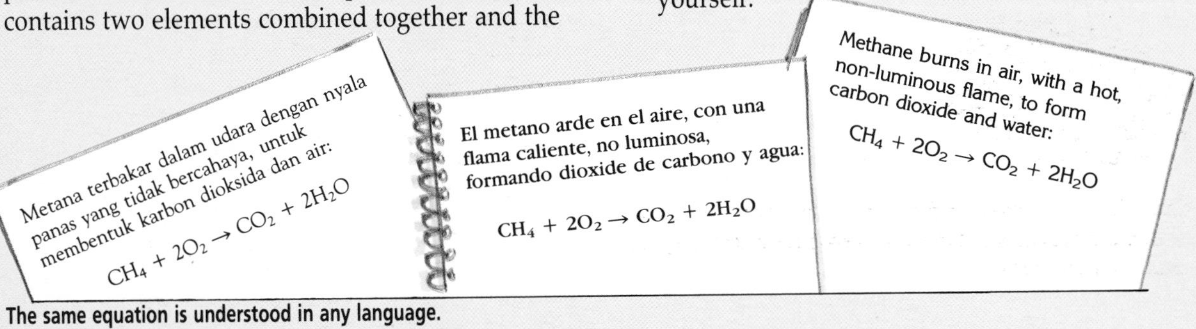

The same equation is understood in any language.

5. How can you use the periodic table?

An obvious way to arrange the elements is in order of the masses of their atoms, putting the element with the least massive atom first.

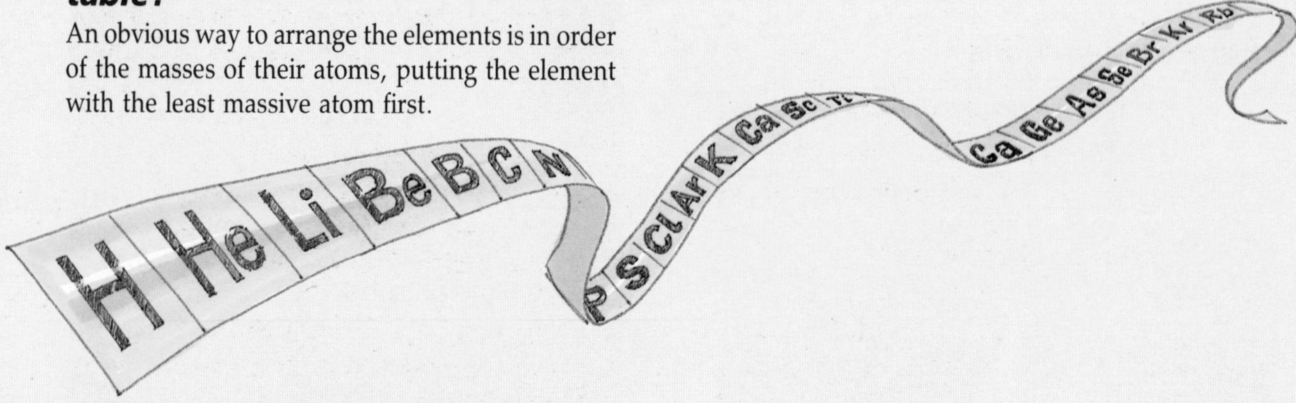

If you do this you might notice that similar elements appear at fairly regular intervals. For example, lithium, sodium and potassium are all soft metals which react with water to form alkaline solutions. The classification can be improved – instead of having the elements in one long row a new row is started every time you come to one of these similar elements.

This arrangement is more compact than the long list. If you look closely at it you will see other examples of where elements in the same vertical column have similar properties.

A more complete table is shown below. The numbers under the symbols show the order of the elements in the table. This number is called the **atomic number** of each element.

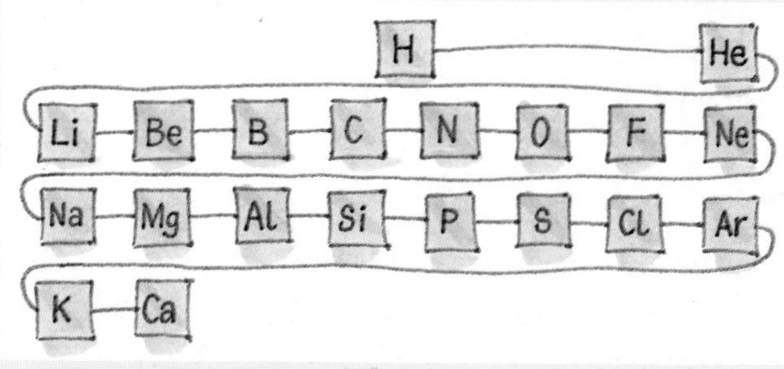

Remember that the table is continuous, but a new row is started at periodic intervals so that similar elements are under one another. The horizontal rows are called periods. The vertical columns which contain similar elements are called groups.

group I	group II											group III	group IV	group V	group VI	group VII	group 0
						H 1											He 2
Li 3	Be 4											B 5	C 6	N 7	O 8	F 9	Ne 10
Na 11	Mg 12											Al 13	Si 14	P 15	S 16	Cl 17	Ar 18
K 19	Ca 20	Sc 21	Ti 22	V 23	Cr 24	Mn 25	Fe 26	Co 27	Ni 28	Cu 29	Zn 30	Ga 31	Ge 32	As 33	Se 34	Br 35	Kr 36
Rb 37	Sr 38	Y 39	Zr 40	Nb 41	Mo 42	Tc 43	Ru 44	Rh 45	Pd 46	Ag 47	Cd 48	In 49	Sn 50	Sb 51	Te 52	I 53	Xe 54
Cs 55	Ba 56	La 57	Hf 72	Ta 73	W 74	Re 75	Os 76	Ir 77	Pt 78	Au 79	Hg 80	Tl 81	Pb 82	Bi 83	Po 84	At 85	Rn 86
Fr 87	Ra 88	Ac 89															

The left-hand block contains the more reactive metals

The central block contains many of the metals which are used to make things

The non-metals sulphur, nitrogen and chlorine, which are all used to make acids, are in the right-hand block

You can link the properties of elements to where the elements are in the periodic table. This will help you to cope with extra information as you learn more about the elements.

Things to do

Transporting Chemicals

Things to try out

1 Baking powder, salt, talcum powder and sugar are four white, solid chemicals which you probably have at home. Like the chemicals mentioned in this chapter, they have different properties. Use the tests below to devise a key which could be used to distinguish between the four substances.

Test 1
Shake a small quantity with water to see if the substance dissolves.

Test 2
If it dissolves, make a more concentrated solution. Test it with a piece of litmus paper.

Test 3
Add vinegar (which is an acid) to each.

Test 4
Line a baking tray with aluminium foil. Place a small heap of each substance at opposite corners of the tray – make sure you can remember which substance is which. Put the tray in an oven set at 200 °C and heat the substances for 15 minutes. Switch off the oven and leave it to cool. Then, using an oven glove, carefully remove the tray and see if the heat has affected the substances differently.

Things to find out

2 Use the periodic table opposite.

(a) Find the position of the element barium. From its position in the table predict

(i) whether it is likely to be a metal or a non-metal

(ii) whether it is likely to react more or less vigorously with water than calcium.

Explain the reasons for your predictions.

(b) Find the position of the element krypton. From its position in the table predict

(i) whether it is likely to be a solid, liquid or a gas

(ii) whether it is likely to be a metal or a non-metal.

If possible, check your predictions by looking up the properties of barium and krypton in an advanced chemistry book.

Things to write about

3 In this chapter the words atom, element, compound and molecule are mentioned.

Write an explanation of what these words represent and how they are related to each other. Aim your explanation at someone who is in the year below you at school. If it helps, use diagrams in your explanation.

Making decisions

4 A tanker has been involved in a road accident. It looks as though some of the chemical it is carrying is leaking out of the tanker. The sign on the back of the tanker is the same as the one below.

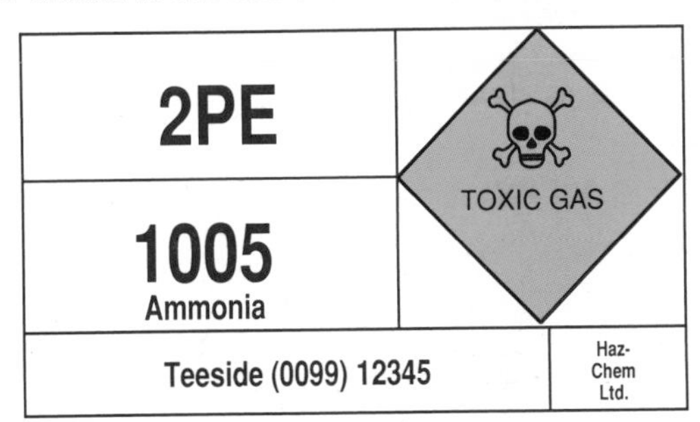

Use the Hazchem code on page 39 to decide what the fire service should do when they arrive at the accident.

5 The Hazchem code for the transportation of chemicals is explained on page 39. There are many chemicals in most homes which are dangerous, for example, cooking oil, bleach, medicines, paintbrush cleaners and weedkillers. Devise a Hazchem code which would be useful in the home. Base it on either the chemicals listed or others which you think are more appropriate.

Points to discuss

6 On your own, try to think of one object used at home which does not use a product of the chemical industry in its manufacture. Then discuss your ideas with each other. If you still think you have thought of an object that does not need the chemical industry, discuss it with your teacher.

7 The formula of vitamin C is $C_6H_8O_6$. What information does this formula tell you about the compound? Vitamin C, whether made artificially from oil or extracted from orange juice, still has the same formula. Think about this and discuss whether or not it is better to buy a drink with added artificial vitamin C.

Questions to answer

8 Look at the ingredients on these labels.

> *Indigestion tablets*
> Each tablet contains calcium carbonate, $CaCO_3$, 680 mg and magnesium carbonate, $MgCO_3$, 80 mg.

> *Cough mixture*
> Each 5 cm^3 contains diphenhydramine hydrochloride, $C_7H_{22}NOCl$, 14 mg.

Make lists of all the different elements in each medicine. Look at the periodic table on page 42 and find out where each element is on the table. Use this to help you to predict whether each element is a metal or non-metal.

Look at *In Brief 5*, page 37. For each ingredient, write out the ratio of the numbers of atoms of each element present in the ingredient.

For questions **9** to **13**:

A SO_2
B C_2H_4O
C $CaCO_3$
D $C_2H_4O_2$
E Cl_2

From the formulas A to E, choose the one which

9 contains the element sulphur

10 contains the element calcium

11 contains only one element

12 represents the greatest number of atoms

13 contains the elements carbon, oxygen and hydrogen in the ratio of 2 : 1 : 4.

14 Write the word equations for the reactions represented by each of the following equations.

(a) $2Cu + O_2 \rightarrow 2CuO$
(b) $2K + Cl_2 \rightarrow 2KCl$
(c) $Zn + 2HCl \rightarrow ZnCl_2 + H_2$
(d) $N_2 + 3H_2 \rightarrow 2NH_3$

15 (a) The following two equations are balanced.

$$2Fe + 3Cl_2 \rightarrow 2FeCl_3$$

$$Ca(OH)_2 + 2HCl \rightarrow CaCl_2 + 2H_2O$$

They are called balanced equations because they have the same number of atoms of each element on each side of the equation.

For each equation count the number of atoms of each element on each side and so show they are balanced.

(b) The following two equations are *not* balanced.

$$CO + O_2 \rightarrow CO_2$$

$$NaOH + H_2SO_4 \rightarrow Na_2SO_4 + H_2O$$

For each equation, count the number of atoms of each element on each side. Then by putting appropriate numbers in front of formulas make the equations balance.

For questions **16** to **20**:

A $CH_4 + 2O_2 \rightarrow CO_2 + 2H_2O$
B $Mg + Cl_2 \rightarrow MgCl_2$
C $H_2SO_4 + Fe \rightarrow FeSO_4 + H_2$
D $H_2 + O_2 \rightarrow H_2O$
E $Fe + CuCl_2 \rightarrow FeCl_2 + Cu$

From the equations A to E, choose the one which

16 represents a reaction which produces hydrogen

17 contains the greatest number of elements

18 represents a reaction between a metallic element and a non-metallic element to form a compound

19 represents the burning of a fuel to form carbon dioxide and water

20 is *not* a balanced equation.

Introducing CONSTRUCTION MATERIALS

What materials is your home made of? Are your school, library, town hall and shops all made of the same materials?

When buildings are designed the architect or engineer has to decide which materials to use. The choice will depend on

- the properties of the materials
- the appearance of the materials
- the cost of the materials.

Look at this building site.

1. Make a list of the materials being used and what you think each is being used for.
2. For each material try to think of one alternative material which could be used for the same purpose. Write the names of these alternative materials on your list.

IN THIS CHAPTER YOU WILL FIND OUT

- about the properties of construction materials
- how these properties are related to the structures of the materials
- how you can sometimes modify the properties of materials.

Looking at

A Continuous Process for Making Glass

Glass is a very important construction material – all buildings have windows, and some look as if they are made completely of glass. So we need to produce glass as quickly and cheaply as possible to keep up supplies.

Batch production

Batches of glass can be made like this. ▶

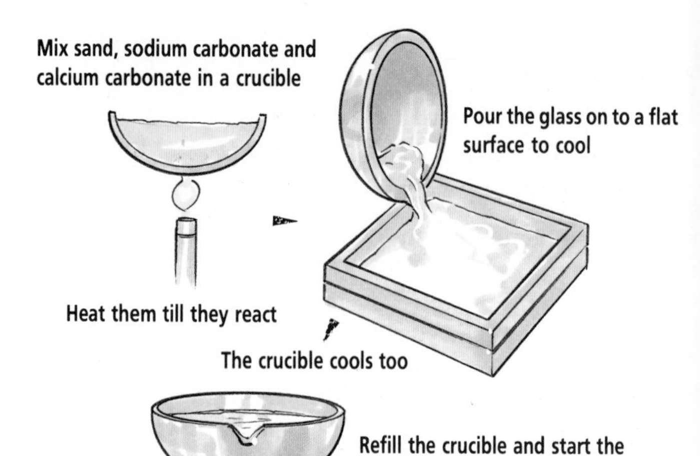

Only small pieces of glass can be made like this, and on an industrial scale lots of people would be involved in the process.

> **1 Do you think this method uses energy efficiently? Where is energy used inefficiently?**

Continuous production

To improve production people have developed a **continuous** process which avoids hold-ups and energy waste. They had to solve a few problems.

Problem 1

How do we get a continuous supply of raw materials into the furnace to be heated?

Solution

Feed them in through hoppers. ▶

Problem 2

How do we get the glass to come out of the furnace continuously?

Solution

A ribbon of molten glass comes out of the furnace. The ribbon is unbroken through the next stages of production, so the cooled glass further along keeps pulling new molten glass out of the furnace.

Problem 3

What can we rest the glass on while it is cooling down? The glass must be kept moving, so it cannot be anything that it will stick to. The glass is very hot, so it cannot be something which will burn.

Solution

Float the glass on molten tin. The tin is very hot – nearly as hot as the glass. This means that the glass cools slowly. ▶

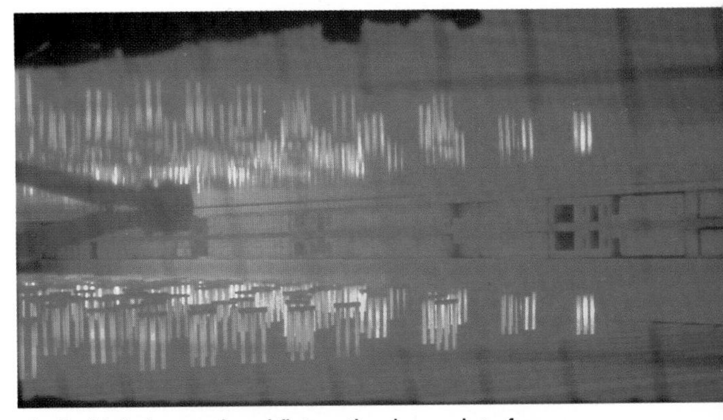

The liquid tin is smooth and flat, so the glass cools to form a smooth, flat surface.

Problem 4

How can we keep very hot solid glass moving?

Solution

Feed the hot ribbon of glass over rollers inside a special oven, where the glass cools down slowly. ▶ When the glass reaches the end of the oven it has cooled to 200°C. The glass is now cool enough to be cut up into sheets.

The glass is cut to order – the sizing and cutting are controlled by computer.

2 Draw a diagram of a float glass production line. Make notes on your diagram to show how making glass this way uses energy more efficiently than traditional methods.

Looking at

Pottery in Buildings

What does the word 'pottery' mean to you? Plates and pots? You may be surprised to find that pottery is also used in the construction of your home, and in the buildings around you.

The tiles on these rooves are made of pottery.

How is pottery made?

Like brick, pottery is made from clay. The difference between them is that pottery is normally heated (**fired**) twice, while brick is only fired once. The higher the temperature the pottery is heated to, the less water it will let through. Pottery surfaces that do not absorb water are extremely useful in bathrooms and kitchens.

Pottery is also easy to decorate. People make it into interesting shapes before firing it, or cover it in special chemicals called **glazes**. Glazes react with the clay during firing to give a shiny and durable coating.

1. Design a model house or other building of your choice. Make a list of places where pottery can be used in it (there are some ideas on this page).
2. Why is pottery chosen for each of the jobs on this page? Add reasons to your list. What other materials could be used instead? Is pottery the only suitable material?

Pottery is used for the wall and floor tiles as well as the pots in this kitchen.

These pottery tiles form cladding on the walls of a house.

The plates are made from pottery as well as the washbasin and toilet.

These are pottery mosaic tiles at the Sheikh Lutfullah Mosque, Iran.

OVERHALL NEEDED!

London's Albert Hall is in danger of falling down. The much-loved London landmark is crumbling away, with water penetrating deeper into the famous stonework each winter. And how has this disaster come about? Now there's the **rub** . . .

In the 1970s the Albert Hall was looking rather dirty. So it was cleaned – by sandblasting. Sandblasting not only rubs off the dirt, it also rubs off the top layer of the building material.

The Albert Hall before sandblasting.

After sandblasting – the Albert Hall is clean but it's also crumbling!

The Albert Hall isn't just any old building. All that elaborate decoration was done in terracotta – a type of pottery. Terracotta has a thin layer of material on the outside which makes it highly resistant to water and very long lasting. But when that layer is rubbed away, the stonework underneath is easily penetrated by water. When this water freezes, it expands. And all those tiny pockets of ice are making the old stones crumble.

If something isn't done soon the ice will be 'bringing the house down'.

3 The owners of the building are extremely worried about what to do. Read through these possibilities and discuss which one you think is the best course of action.
   (a) The building could be left to crumble and collapse.
   (b) The outside could be completely rebuilt in terracotta, which would be very expensive.
   (c) The building could be treated with liquid polymers. These are used to stop water getting into high-rise blocks built of concrete – they soak into the material and form a water-resistant layer. However, using these polymers might be a risk, because no one knows how they will react with terracotta over a long period of time. Once they are applied, the process cannot be reversed.
   (d) The building could be treated with silicone waxes. These act like other waxes and polishes – they form a water-resistant layer. However, unlike liquid polymers, they need to be replaced after about five years, as they are gradually washed away by rain.
4 Write a report for the owners explaining why you reached the conclusion you did, and why you rejected the other three solutions.
5 Imagine that it is one year later and your report has been acted on. Write a follow-up article to the one above, describing what has happened.

Looking at

The Heights of Buildings

One of the laws of Hammurabi, King of Babylonia around 2200 BC, reads:

'If a builder builds a house that is not firmly constructed and it collapses and causes the death of its owner, that builder shall be put to death. If the collapse causes the death of the son of the owner, the son of the builder shall be put to death.'

Even without such penalties, builders want their buildings to last. The sort of building that can be built depends on the strengths of the materials available. As new stronger materials have been developed, so builders have built higher and higher buildings.

These pictures show the tallest structures in the world built of a number of different materials.

Town Hall, Siena, Italy, 102 m, built of brick in 1348.

Notre Dame Cathedral, Paris, France, 141 m, built of sandstone in 1439.

Eiffel Tower, Paris, France, 300 m, built of iron in 1889.

The Pyramids, El Gizeh, Egypt, 147 m, built of limestone in 2580 BC.

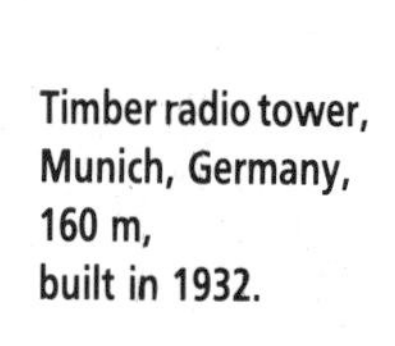

Timber radio tower, Munich, Germany, 160 m, built in 1932.

Canadian National Tower, Toronto, Canada, 555 m, built of concrete in 1975.

Steel radio mast, Plock, Poland, 646 m, built 1974.

Washington Memorial, Washington, USA, 169 m, built of stone in 1884.

1 Make a list of these buildings in the order of when they were built. Start with the one which was built first. What materials have the highest buildings been made of? Have these materials always been available?

2 Use the information on this page to draw a diagram showing how the materials used have influenced the heights of buildings. Draw each building to scale (so if you decide to use a scale of 1 cm = 50 m, the Eiffel Tower drawing would be 6 cm high). Label each drawing with the type of construction material used.

In brief

Construction Materials

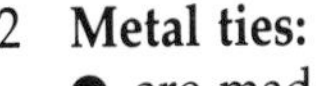

1 Bricks:
- are made from fired clay
- are strong and do not absorb much water.
- The atoms in clay are bonded together in layers. During firing the layers become bonded together to form a three-dimensional structure.

2 Metal ties:
- are made from galvanized iron.
- Galvanizing is one way of preventing rusting.
- Both water and oxygen are needed for rusting.

3 Plastics:
- are made by polymerizing certain compounds obtained from crude oil
- are light and weather resistant.
- Some can be reshaped when heated – **thermoplastics**.
- Some cannot be reshaped once formed – **thermosets**.
- During polymerization, large numbers of small molecules (**monomers**) combine to form a very large molecule (**polymer**).

4 Aluminium:
- is resistant to corrosion because of the layer of aluminium oxide on its surface.

5 Glass:
- is made by heating sand, soda ash and limestone together
- has a three-dimensional structure and is hard. But, like a liquid, it does not have a regular structure
- can be made safer by heat-toughening or laminating with plastic.

7 Paint:
- is a mixture of a polymer, a solvent and a pigment
- is used to protect wood from rotting and metal from corroding
- sets when the solvent evaporates and the polymer chains react with oxygen from the atmosphere to bond together.

6 Wood:
- is hard, but can be split along the grain and will absorb water and rot
- has long thin molecules lined up along the grain with weak bonds between them.

Thinking about

Construction Materials

1. Why do different construction materials have different properties?

You choose a particular material because it has the best properties for the job you have in mind. You can use theories about the *structures* of materials to explain their *properties*.

Everything consists of tiny particles which are too small to see. How the particles in a material are arranged and held together is called the **structure** of the material.

Obvious differences between construction materials can be explained by classifying them as one-, two- or three-dimensional structures.

Structures consisting of long, thin particles (one-dimensional structures) explain some properties of wood, wet paint and some plastics.

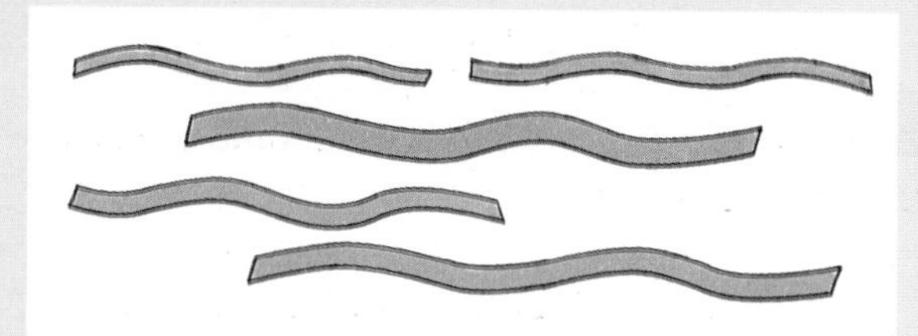

The particles (molecules) are not held strongly together. This means that:

Wet paint — is flexible and runny;

Wood — can be split along its grain.

Graphite (a form of carbon) has particles held together in large flat molecules (a two-dimensional structure). Clay and slate have a similar structure.

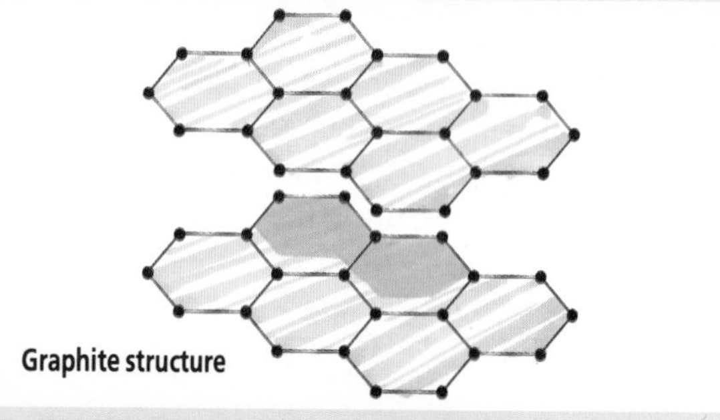

Graphite structure

The flat molecules can slide over each other. This means that:

Clay — is pliable and slippery when wet;

Slate — can be split into sheets.

Diamond (another form of carbon) has particles arranged in a big interlocking structure (a three-dimensional structure). *Hard* materials like brick, stone, concrete, glass, dry paint and some plastics have a similar structure.

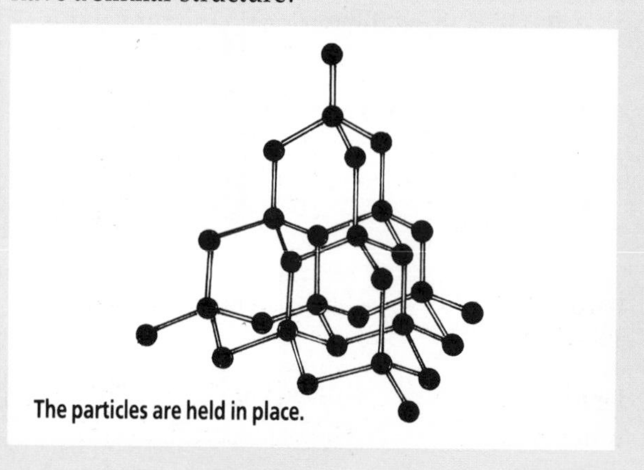

The particles are held in place.

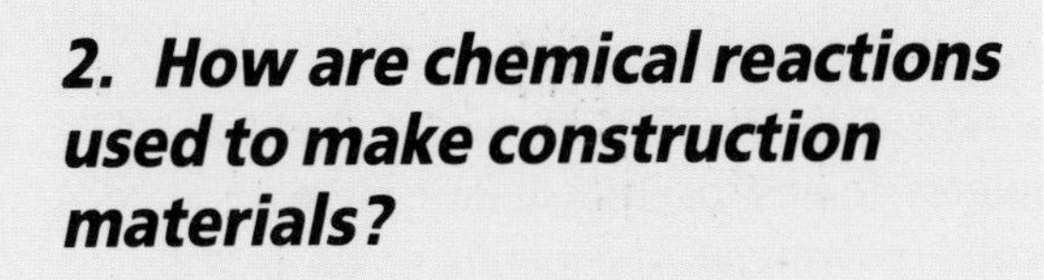

2. How are chemical reactions used to make construction materials?

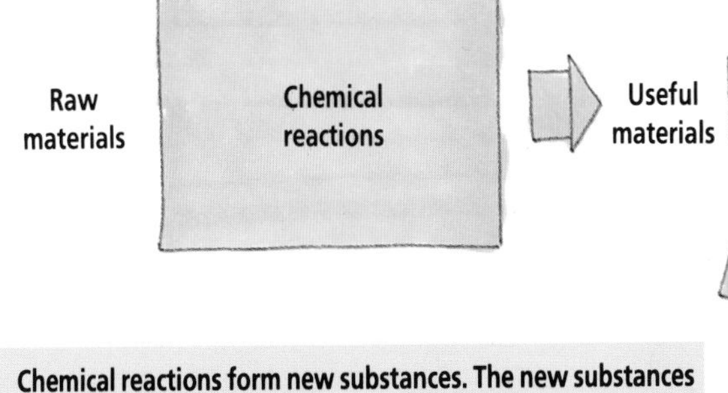

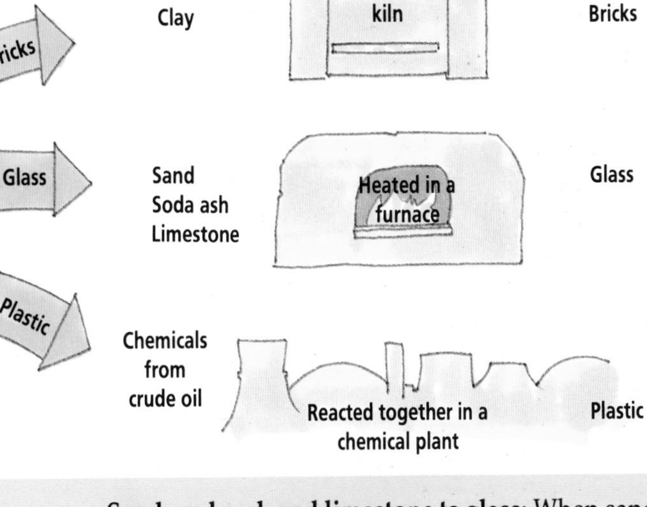

Chemical reactions form new substances. The new substances here are brick, glass and plastic. These substances have more useful properties than those they were made from.

3. What happens during these chemical reactions?

Clay to bricks: Clay contains aluminium, silicon and oxygen atoms bonded together in separate flat layers. When the clay is wet, water molecules get between the layers and allow them to slide over one another.

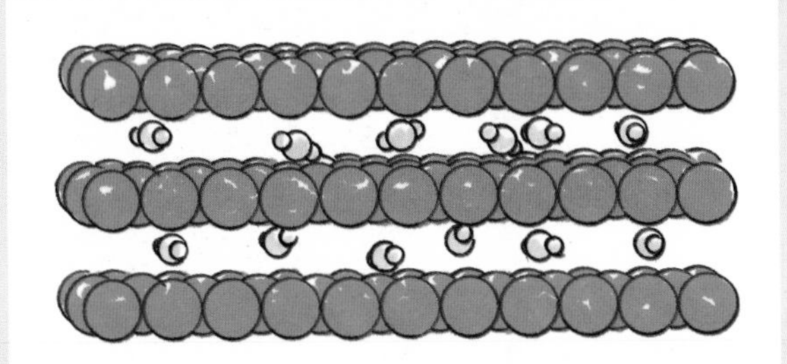

Firing clay (heating it strongly) drives out all the water molecules. Then atoms in one layer form bonds with atoms in another.

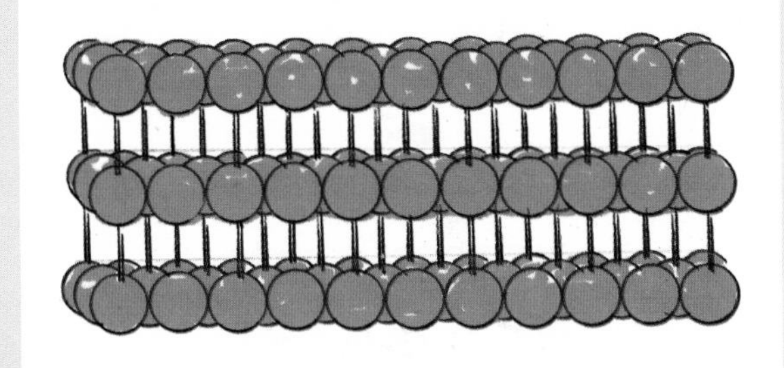

This cross-linking gives brick a three-dimensional structure and makes it hard.

Fired clay cannot be converted back into clay when water is added. So your house is safe in the rain!

Sand, soda ash and limestone to glass: When sand (silicon dioxide) is heated to a very high temperature, it melts and then forms a clear glassy solid when cool.

If sand is heated with soda ash (sodium carbonate), the mixture melts at a lower temperature. This saves energy, but unfortunately the product is soluble in water and so is no use as a glass.

If limestone (calcium carbonate) is also added to the mixture before heating, it still takes less energy to make the glass. The glass formed this time is *not* soluble in water. Normal glass is made by heating these three raw materials together.

Taking it further

When the substances used to make glass are heated together they melt. But there is more to it than that. New substances are formed so chemical reactions have taken place.

Sand is silicon dioxide (SiO_2).

Soda ash is sodium carbonate (Na_2CO_3).

Limestone is calcium carbonate ($CaCO_3$).

The reactions which occur are:

silicon dioxide + **sodium carbonate** → **sodium silicate** + **carbon dioxide**

silicon dioxide + **calcium carbonate** → **calcium silicate** + **carbon dioxide**

Ordinary glass is a mixture of sodium and calcium silicates.

Why is glass different? Glass is unusual. It is hard but unlike most solids it does not have a sharp melting point.

Glass can be very tough — it can be reinforced.

Molten glass is runny — it can be poured.

In most solids the particles are in a regular three-dimensional arrangement. In glass, the particles are in a three-dimensional structure but the arrangement is not regular. Glass is really a liquid which has been cooled so quickly that the particles have not had time to organise themselves into a regular pattern. Like a liquid, glass is transparent.

Chemicals from oil to plastic

Plastics are made from lots of small molecules (**monomers**) joined together to form large molecules (**polymers**). The joining up process is called **polymerization**. The monomers are usually made from crude oil.

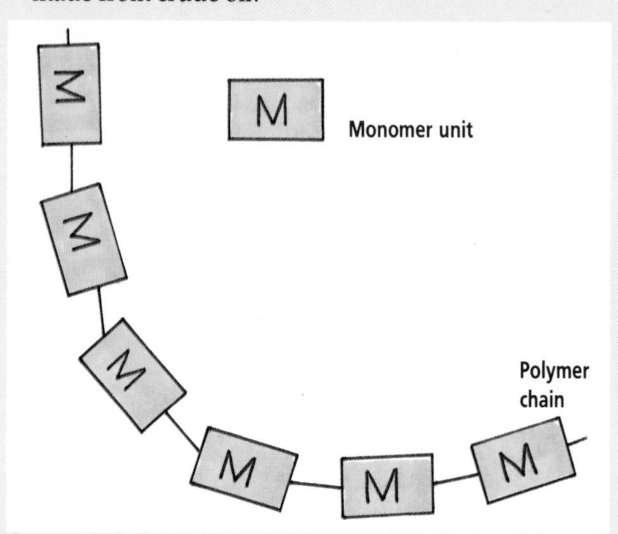

This polymer is made from only one type of monomer.

Adding monomers together like this is called **addition polymerization**. Sometimes monomers combine by a smaller molecule being 'condensed out'. This is called **condensation polymerization**. Many of these polymers are formed from a mixture of two different monomers.

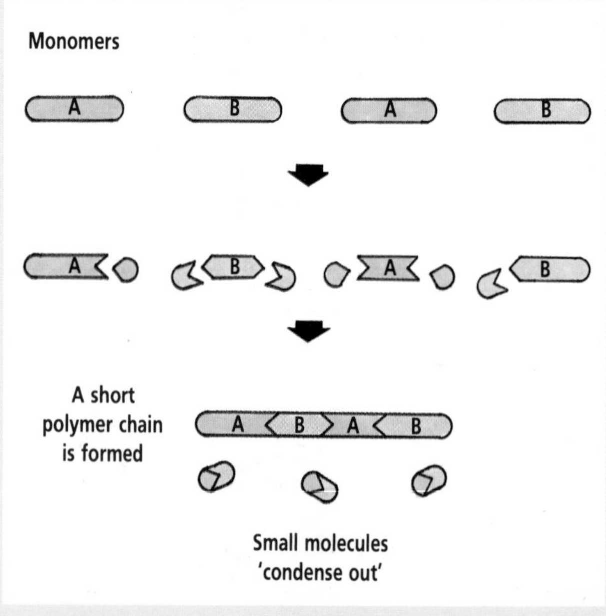

This condensation polymer is made up from two types of monomer.

Sometimes the monomers are gases like ethene or liquids like chloroethene. When they polymerize they become solids. Poly(ethene) (polythene) and poly(chloroethene) are solids. (Chloroethene used to be called vinyl chloride, so the polymer was called poly(vinylchloride) – PVC).

Combining lots of little molecules together to form big molecules has converted the gas or liquid into a solid. Substances made up of little molecules are more likely to be gases or liquids at room temperature. Substances made up of big molecules are more likely to be solids.

Some polymer molecules are long and thin, almost one-dimensional. Then the plastic is flexible. If it is heated it melts and can be reshaped. It is a **thermoplastic**.

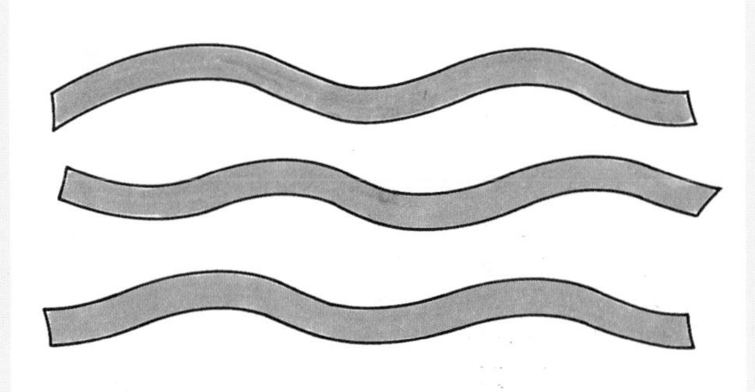

Thermoplastics have no cross-links between their molecules.

Some condensation polymerizations form polymers with three-dimensional structures.

Plastics made from these polymers are more rigid and they do not melt so easily. When they are heated they decompose. These are called **thermosetting plastics** because once set in shape they cannot be reshaped by heating.

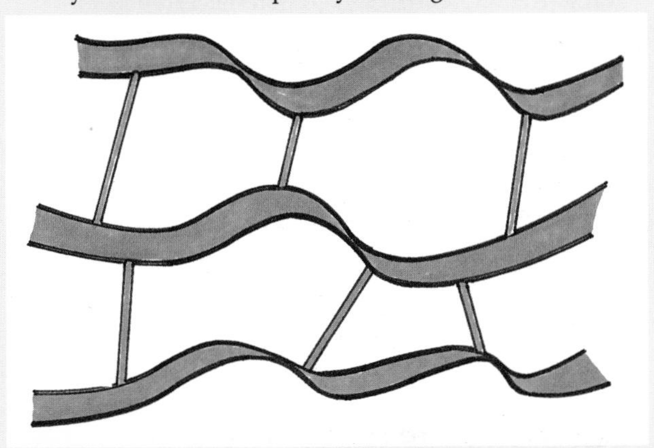

Thermosetting plastics have cross-links between their molecules.

4. What happens when oil paint sets?

There are three main ingredients in paint.

When paint dries two things happen:

- the solvent evaporates (changes to a gas). That is why wet paint has such a strong smell
- wet paint has a one-dimensional structure like that of thermoplastics. Oxygen from the atmosphere reacts with the paint molecules and forms cross-links between them. This gives a three-dimensional structure like that of thermosetting plastics which is much harder.

5. What happens when metals corrode?

When paint is chipped off a car or bicycle frame the iron underneath gets wet. Wet iron begins to **rust**. Rusting is a special name for the corrosion of iron.

Iron only rusts if the surface of the iron is in contact with the air *and* with moisture.

Although water is necessary for rusting, the reaction can be represented by:

iron + oxygen → iron oxide

$$4Fe + 3O_2 \rightarrow 2Fe_2O_3$$

Rust is a form of iron oxide.

Dry rust weighs more than the original iron because it has captured some oxygen from the air. When oxygen is added to another substance the process is called **oxidation.**

Whenever a metal corrodes, oxidation occurs.

When silver goes black or a freshly cut piece of aluminium loses its shine, the same sort of chemical changes are happening. The metal is being attacked (corroded) by the oxygen in the air.

aluminium + oxygen → aluminium oxide

$$4Al + 3O_2 \rightarrow 2Al_2O_3$$

But silver and aluminium do not corrode away completely. Silver is not very reactive so it corrodes very slowly.

Aluminium is much more reactive than silver, but as soon as some aluminium oxide is formed on the surface of the metal it protects the rest of the metal. This protective layer makes aluminium very useful. Aluminium door frames do not need painting.

6. How quickly does an acid react with limestone?

Limestone is a very popular building stone. Gases are released into the atmosphere by power stations which burn fossil fuels. Some of these gases can make the rain slightly acidic. This **acid rain** attacks limestone buildings.

Limestone is calcium carbonate. When it reacts with an acid, carbon dioxide gas is given off. For example:

calcium carbonate ($CaCO_3$) + **hydrochloric acid** ($2HCl$)

↓

calcium chloride ($CaCl_2$) + **water** (H_2O) + **carbon dioxide** (CO_2)

You can investigate this reaction by measuring the volume of gas given off at regular time intervals. You can plot the results on a graph.

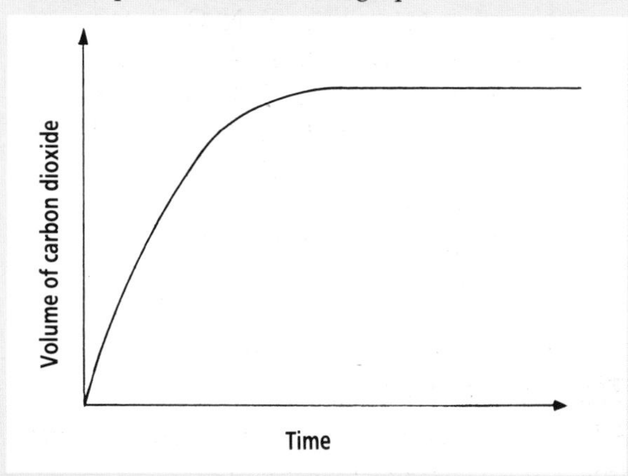

The steeper the curve the quicker the reaction is going (the higher its **rate**). When the curve is flat the reaction has stopped.

If you investigated the reaction in this way you would find that there are three main ways to speed up the reaction:

(a) using small pieces of limestone rather than big lumps. This increases the surface area of the limestone so the acid can attack the stone more quickly

(b) heating the acid. This gives the particles of acid more energy

(c) using more concentrated acid. This means there are more acid particles around to collide with the surface of the limestone.

Things to do

Construction Materials

Things to try out

1 Bricks are slightly porous. So they can absorb water. The amount they absorb can be expressed as 'the percentage increase in their mass when they are placed in water'.

$$\text{water absorption} = \frac{\text{increase in mass}}{\text{mass of dry brick}} \times 100\%$$

Plan how you would compare the amount of water absorbed by different bricks.

Find some different bricks and try out your method.

2 The names of ten building materials are hidden in the wordsearch below. How many can you find?

<table>
<tr><td>F</td><td>A</td><td>Y</td><td>J</td><td>E</td><td>M</td><td>I</td><td>O</td><td>D</td><td>G</td><td>E</td><td>D</td></tr>
<tr><td>I</td><td>B</td><td>P</td><td>O</td><td>S</td><td>T</td><td>E</td><td>E</td><td>L</td><td>N</td><td>R</td><td>B</td></tr>
<tr><td>K</td><td>P</td><td>E</td><td>D</td><td>C</td><td>O</td><td>H</td><td>T</td><td>U</td><td>L</td><td>A</td><td>I</td></tr>
<tr><td>C</td><td>A</td><td>L</td><td>M</td><td>I</td><td>A</td><td>L</td><td>E</td><td>N</td><td>O</td><td>T</td><td>S</td></tr>
<tr><td>H</td><td>G</td><td>I</td><td>A</td><td>W</td><td>E</td><td>B</td><td>R</td><td>A</td><td>E</td><td>K</td><td>C</td></tr>
<tr><td>G</td><td>L</td><td>A</td><td>S</td><td>S</td><td>D</td><td>O</td><td>C</td><td>F</td><td>O</td><td>Y</td><td>G</td></tr>
<tr><td>N</td><td>I</td><td>M</td><td>U</td><td>F</td><td>T</td><td>E</td><td>N</td><td>W</td><td>A</td><td>R</td><td>M</td></tr>
<tr><td>A</td><td>P</td><td>O</td><td>C</td><td>I</td><td>J</td><td>I</td><td>O</td><td>L</td><td>E</td><td>T</td><td>U</td></tr>
<tr><td>E</td><td>N</td><td>H</td><td>M</td><td>B</td><td>A</td><td>M</td><td>C</td><td>C</td><td>O</td><td>G</td><td>D</td></tr>
<tr><td>L</td><td>U</td><td>B</td><td>G</td><td>E</td><td>V</td><td>A</td><td>N</td><td>S</td><td>H</td><td>D</td><td>A</td></tr>
<tr><td>F</td><td>E</td><td>O</td><td>K</td><td>C</td><td>I</td><td>P</td><td>L</td><td>V</td><td>B</td><td>J</td><td>L</td></tr>
<tr><td>R</td><td>C</td><td>R</td><td>E</td><td>D</td><td>W</td><td>T</td><td>H</td><td>A</td><td>T</td><td>C</td><td>H</td></tr>
</table>

Things to find out

3 Because bricks are porous they can absorb water from the ground or from driving rain.

(a) How do builders prevent buildings absorbing water from the ground?

(b) Why does driving rain cause greater problems in old buildings which do not have cavity walls?

(c) What problems result when the brickwork of a house is permanently damp?

4 (a) What are the commonest types of rock in your area?

(b) Are there any quarries (open or disused) in your area? If so what do (did) they produce?

(c) What type of stone is used in your local buildings?

5 Colourless glass is made by heating sand, soda ash and limestone together. Coloured glass is made by adding small amounts of other substances. Find out what substances are added to make it **(a)** blue **(b)** green **(c)** red.

Amber Palace, India

Points to discuss

6 All construction materials use natural resources. For example, window frames can be made from wood, steel, aluminium or plastic.

Wood comes from trees, steel and aluminium are extracted from minerals, plastic is made from oil. All these processes which convert raw materials into window frames use energy. Making aluminium and plastic uses most energy, making steel uses less and processing wood uses much less.

Discuss which material should be used for window frames in new houses. Some properties of these materials may make them more suitable. How does this balance against the need to conserve resources?

Questions to answer

7 Read *Thinking about 1* (page 52) and use the ideas to answer the following questions:

(a) Why do you think slate is easier to split than limestone?

(b) Carrier bags and the casings of electric plugs are made from different plastics. Make a list of the ways in which the two plastics are similar and the ways in which they are different. How would you expect the structures of the two plastics to differ?

8

(a) Give the scientific explanation of the difference between addition and condensation polymers.

(b) Look at page 54. If monomer A is:

and monomer B is:

what would be the formula and name of the small molecule produced during the polymerization?

9 This graph shows the rate of reaction between some limestone chips and 1.0 M hydrochloric acid at 25°C.

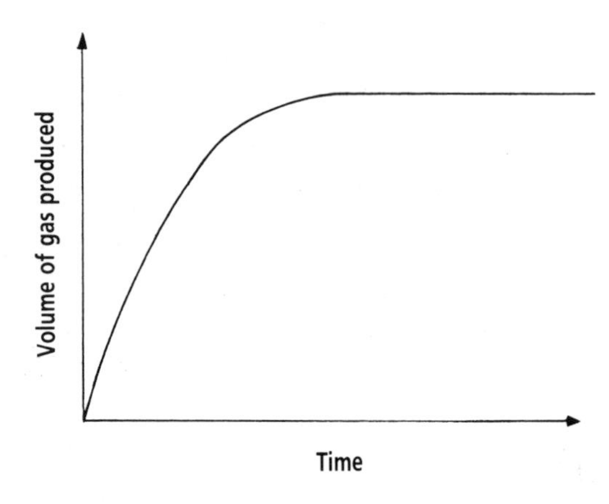

Redraw the graph and sketch in the lines you would expect if:

(a) the same mass of limestone *powder* was used. Label this line A.

(b) the temperature was 15 °C. Label this line B.

(c) half the mass of limestone chips was used. Label this line C.

10 This question is about clay and bricks.

(a) Why is clay slippery and smooth when it is wet?

(b) What happens to the structure of clay when it is changed into brick?

(c) Why are bricks hard and rigid?

(d) How does frost damage bricks?

11 When damp iron rusts it looks as if it is being 'eaten away', but in fact it is combining with oxygen from the air.

(a) Plan an investigation to show that iron gains something from the air when it rusts.

(b) Aluminium is a more reactive metal than iron. Yet when it is used to make window frames it resists corrosion better than iron.

(i) Explain why aluminium is more resistant to corrosion than iron.

(ii) Describe and explain two ways to prevent the corrosion (rusting) of iron.

Introducing

MOVING ON

How did you and your friends get to school this morning? You probably used various forms of transport between you.

1 Make a list of all the different kinds of transport you have used in the past year. Add the advantages and disadvantages of each kind.

Which forms of transport do you think are safest? Although many people feel nervous before flying, it is actually one of the safest forms of transport! Cyclists are involved in far more accidents than aeroplane passengers.

Lots of us use road transport, but there are often accidents on roads. There are many things we can do to make road journeys safer.

2 List all the things you can see in these photos which make the various road users safer.

IN THIS CHAPTER YOU WILL FIND OUT

- how forces are involved in changing the motion of objects
- how acceleration is measured and what causes it
- how we can use our knowledge of forces to make travelling safer
- how we can 'spread' forces over longer times and larger areas, so that they do less damage.

Looking at

How to Make Cars Safer

If you have ever been involved in a road accident you will know how frightening it can be. Even if your car is travelling quite slowly, you still feel quite a jolt when it bumps into a solid obstacle. And everything seems to happen so quickly!

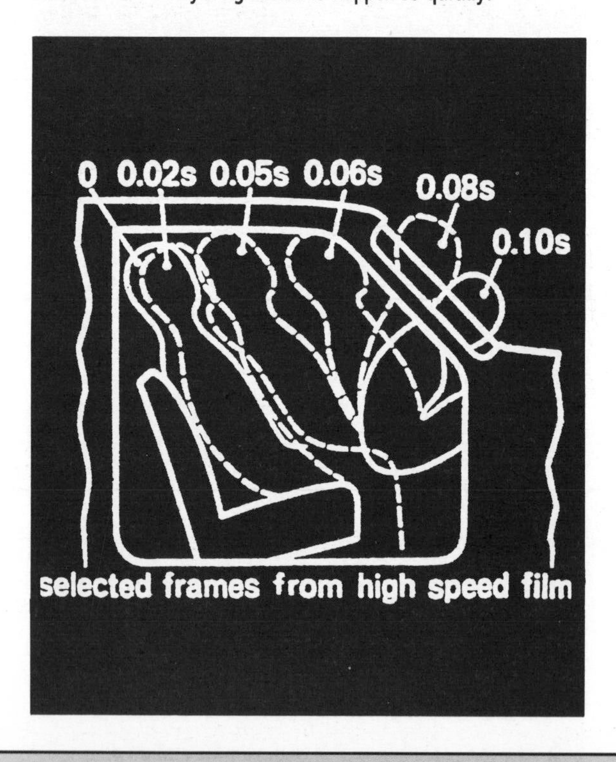

What really happens in a crash?

At the Motor Industry Research Association (MIRA), cars and car safety devices are tested. In some of these tests, cars with dummy passengers are deliberately crashed into a solid wall. A high-speed camera takes a series of photos during the crash. These help researchers to study the movement of the car and its dummy passengers.

Here are the results of one test. The time starts from the moment the front of the bumper touches the wall.

Time (seconds)	What happened
0.026	bumpers pressed in
0.044	driver hits steering wheel
0.05	force on car is at its maximum; bonnet begins to crumple
0.075	driver's head hits windscreen
0.1	driver probably dead
0.11	bonnet completely crumpled;
	rear passenger hurtles into driver's back
0.15	car completely wrecked
0.2	all movements stop

1 On a sheet of paper, draw a time line. Mark what happened in the test at the appropriate times.

0 — 0.1 — 0.2

Time (s)

2 The average person's reaction time is 0.2 second. The fastest possible reaction time is more than 0.1 second. Could the driver react quickly enough to avoid hitting the windscreen? Could the passenger react quickly enough to avoid being thrown forwards into the driver's back?

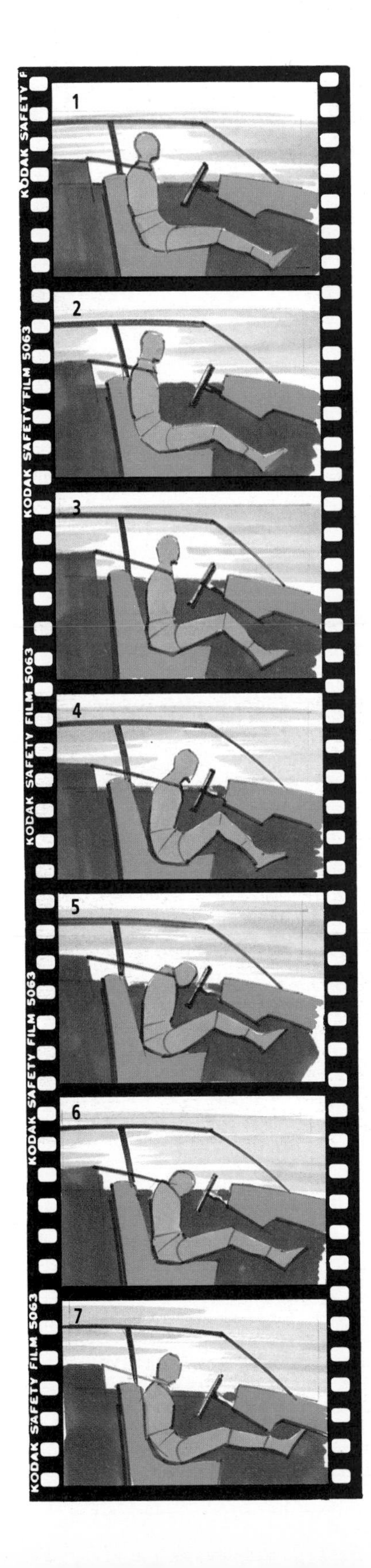

'Belt up': why it makes sense!

The answer to both parts of question 2 is no. This is why seatbelts are such an important safety feature in modern cars. MIRA also test seatbelts. The strip of photos shows how a dummy driver wearing a seatbelt moves during a crash. In photo 1, the crash is just beginning. The time interval between one photo and the next is 1/64 second (0.0156 s).

3 The driver is stationary again in photo 7. How long has the whole crash lasted?

4 Does the driver hit the steering wheel? Does the driver's head hit the windscreen?

5 Using the photos, estimate how long it takes for the seatbelt to stop the driver's body moving forwards. (*Hint:* Count how many photos it takes for the body to stop moving forward and then use the time interval between photos to work out the total time.)

6 Some people think a seatbelt works by holding you completely still. What do the photos show?

How big are the forces?

From the strip of photos, we can also work out how big a force the belt has to exert. The car in these photos was travelling at 80 km/hour (22 m/s) before the crash. This, of course, means that the driver's body was also travelling at 22 m/s before the crash. The driver has a mass of 75 kg. The belt has to slow the driver's body from 22 m/s to zero.

The equation which links these quantities together is:

$$F\,t = mv_{\text{final}} - mv_{\text{initial}}$$

7 Your answer to question 5 is t – the time it takes for the driver's body to stop. v_{final} is zero. Use the equation to work out how big the force F is.

You should find that the force is around 26 000 N. This is almost three times the weight of a small car!

The seatbelt has to be strong enough to exert a force of over 26 000 N. Seatbelts are designed to stretch a little so the body takes a little longer to stop moving. This makes t in the equation larger. This means that F is smaller. If this belt had stopped the driver in just 0.1 s, then F would have had to be over 52 000 N – twice as big.

So there are two reasons why you need to wear a seatbelt:

- You can't react quickly enough to stop yourself, and
- the force you would have to exert with your arms is much bigger than you could possibly manage yourself!

Looking at

Theories about Motion

1 Aristotle's theory says that solid objects fall because they 'naturally' return to their proper place – the Earth. Does this theory agree with what we observe? Could the theory be tested by experiment? Is it a scientific theory?

Predictions based on Aristotle's theory

According to Aristotle, the heavier an object is, the greater its tendency to move towards the Earth. So heavy objects should fall faster than light ones. In fact, for two objects, one twice as heavy as the other, the heavy one should fall to the ground in half the time the lighter one takes.

2 Is this prediction based on everyday experience? Do heavy objects fall faster than light ones? Could this prediction be tested by experiment?

Aristotle's ideas were believed for about 2000 years until . . .

Galileo showed that motion followed the mathematical law he had predicted.

3 **How accurate do you think Galileo's methods would have been? Try to think of another way he might have measured time.**

4 **Did Galileo begin by making observations or by thinking of a theory? What were his experiments for — just to see what happens, or to test a theory?**

5 **Galileo's results agreed with his predictions. Does this mean that his theories about motion are correct? If he had not found what he expected, would this have proved his theories wrong?**

Experimental evidence

Here is a simple way to show that the distance an object falls from rest increases as the **square** of the time.

Take two pieces of thread, just over 1 metre long. Tie five buttons to each thread at the positions shown. Then hold each thread above a table, with the end just touching the table top. Drop one thread, then the other, and listen carefully each time. You should find that the sounds of the buttons landing are equally spaced with thread 1, but not with thread 2.

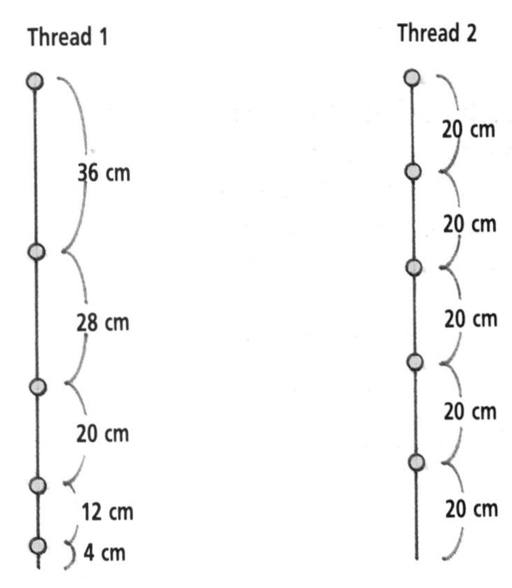

6 **How does the result of this simple experiment show that Galileo's prediction was correct?**

The importance of Galileo's work

Galileo's work was important for two reasons.

- He challenged Aristotle's theory of motion which had stood unquestioned for almost 2000 years. This started people thinking about motion afresh.
- He linked science and mathematics by looking for a mathematical formula to describe the motion of falling objects.

Galileo was also interested in astronomy. Using a telescope he had invented, he discovered that Jupiter had moons. He also observed mountains on the Moon and sunspots. Strange as it may now seem to us, these discoveries led him into great trouble with the Church authorities. They believed that the Earth was the centre of the universe and that the 'heavens' were perfect. Mountains on the Moon, spots on the Sun and moons revolving round something other than the Earth were difficult for them to accept. Galileo was forced to withdraw his opinions and was kept for many years under house arrest.

Looking at

Cycle Helmets

Everything from cost to the 'wally factor' is deterring cyclists from wearing safety headgear. But helmets can save lives

EVERY year, about 300 cyclists are killed on Britain's roads and 5,000 are seriously injured. Two-thirds of these deaths are caused by head injuries.

Many of these deaths and injuries could be prevented by the use of cycling helmets.

In some accidents, the cyclist's head might hit something sharp like the edge of a kerb. So the design of some helmets is tested by putting them on a metal 'head' and dropping a metal spike on to them. If the spike goes right through, it makes an electrical contact and sets off an alarm.

3 Try to sketch an electric circuit diagram showing how this would work.

This cycle helmet is just a single piece of foam polystyrene, with a chin strap! You have probably come across foam polystyrene as a packaging material.

1 Make a list of the properties of foam polystyrene which make it a suitable material to use for a cycle helmet.

4 Cycle helmets could save lives. But they are expensive and some people think they look silly wearing them – the 'wally factor'. Think up a good caption to make the photo on this page into a poster to encourage people to buy and wear cycle helmets.

A kilogram mass has been dropped on this piece of foam polystyrene from a height of 50 cm. What has the impact done to the foam polystyrene?

2 Use this idea to write a short explanation of how the foam polystyrene helmet protects your head in an accident.

In brief

Moving On

1 If an object is at rest (not moving), all the forces on it must be **balanced**. They add to zero.

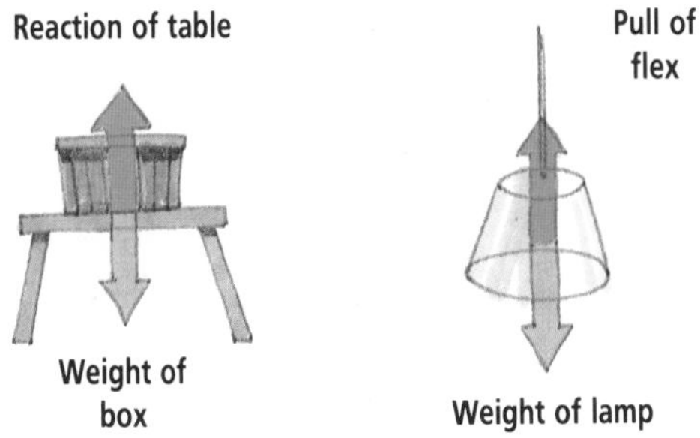

2 Forces **change** motion. Despite what you might think from everyday experience, a force is *not* needed to keep an object moving at a steady speed. This is called **Newton's first law of motion.**

Steady speed, so driving force = counter force

3 A force *is* needed:
- to start an object moving
- to stop an object moving
- to make an object move faster
- to make an object move slower
- to make an object change its direction of motion.

But a force is *not* needed:
- to keep an object moving in a straight line at a steady speed.

4 Left to themselves, objects do not change their motion. This property of all objects is called **inertia.**

5 If a force acts on an object, it makes the object **accelerate.**

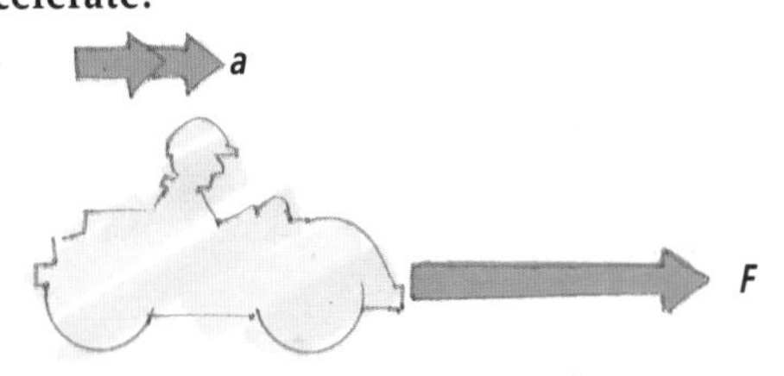

Force causes acceleration

6 Acceleration is defined by the equation:

$$\textbf{acceleration} = \frac{\textbf{change of velocity}}{\textbf{time taken}}$$

The units of acceleration are m.p.h. per second, or m/s per second.

A positive acceleration means that an object is speeding up. A negative acceleration means that it is slowing down.

7 The acceleration of an object depends on the size of the force acting on it and on its mass:

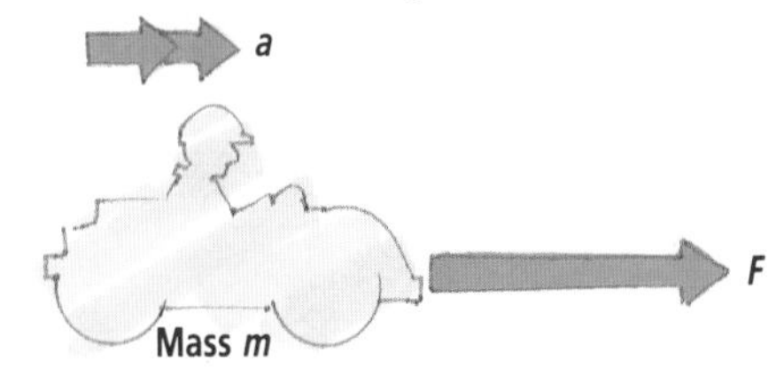

- Acceleration is directly proportional to the force acting.
- Acceleration is inversely proportional to the mass of the object.

This is summarised by the equation:

$$\textbf{force} = \textbf{mass} \times \textbf{acceleration}$$
$$F = ma$$

This is known as **Newton's second law of motion.**

8 The unit of force is the newton (N). One newton is the force needed to give a mass of 1 kg an acceleration of 1 m/s per second.

9 Another useful way to state Newton's second law is:

$$\textbf{force} \times \textbf{time} = \textbf{mass} \times \textbf{change in velocity}$$
$$Ft = mv_{\text{final}} - mv_{\text{initial}}$$

From this it follows that:
- if an object is stopped very quickly, the force involved is large
- if an object is stopped in a longer time, the force involved is smaller.

10 When a force acts on a surface, it exerts a **pressure.** The size of the pressure depends on the force and the area on which it acts:

$$\textbf{pressure} = \frac{\textbf{force}}{\textbf{area}}$$

11 The quantity mv is called **momentum.** When two objects collide, or spring apart (an explosion), the total momentum is the same afterwards as it was before.

Thinking about

Moving On

1. How do forces keep things still?

If you drop an apple, it falls to the floor. A force pulls it downwards – the force of gravity, its **weight.** But if you put the apple on a table, it does not fall – it sits still. Gravity is still pulling the apple downwards but now another force is balancing this out. The table exerts an upward push on the apple, holding it steady.

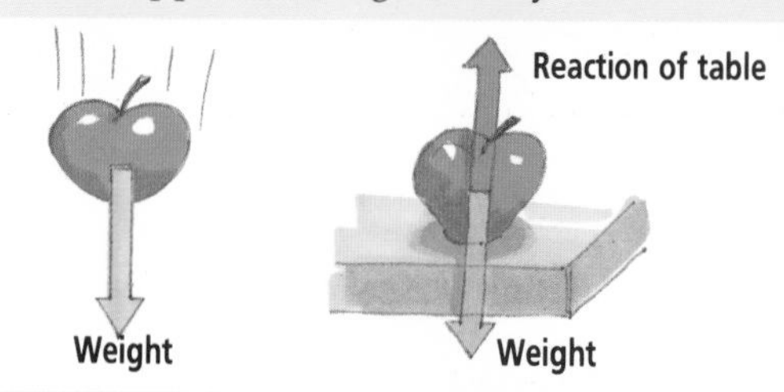

How does the table exert an upwards push? Think what would happen if you put the apple on a piece of foam. The foam would be squeezed down a bit. Foam is springy, so the compressed foam would then exert a push upwards on the apple. The apple sinks in until the upward push is exactly equal to its weight.

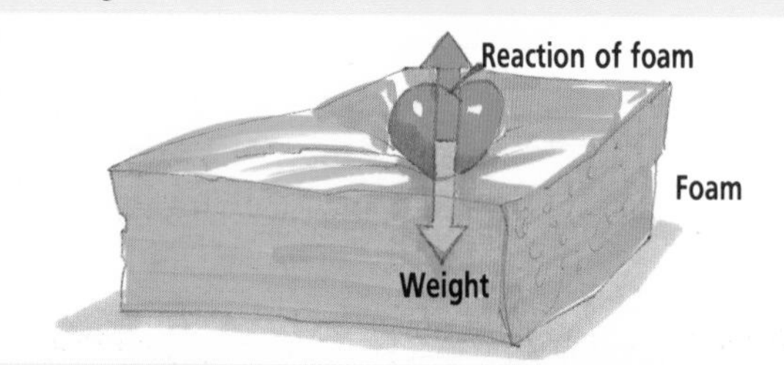

The same thing happens with the apple on the table. The table top is not so easily squeezed as the foam. It *can* be compressed – though not enough to see with the naked eye. The apple 'sinks in' until the upward force balances its weight.

Anything which is sitting or hanging at rest is being held there by **balanced forces.**

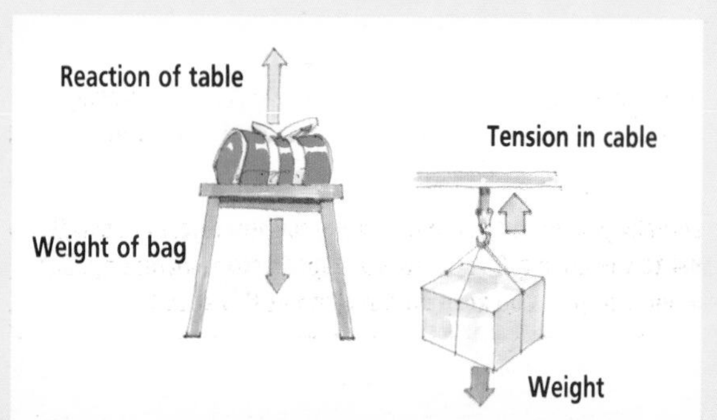

2. Are forces needed to keep things moving?

If you want to move a box across the floor, you have to give it a push or a pull. When you stop pushing or pulling, it stops. It seems that a continuous force (a push or a pull) is needed to make something move. But there are problems with this simple explanation. Things often carry on moving after a force has stopped acting on them.

When you play tennis, your racket hits the ball and exerts a force on it. But the ball keeps on moving after the racket has hit it – even though the racket is no longer exerting any force on it.

Curling is like bowls on ice! The heavy curling stone is set in motion by a push from the player's hand – but then it keeps on moving as it slides down the ice rink towards its target.

The above photo gives us an important clue. The stone keeps on moving because ice is very smooth. There is very little **friction.** If we could get rid of friction altogether, anything that had been set in motion would carry on moving at a steady speed.

Without friction, no force is needed to keep an object moving at a steady speed in a straight line.

Inertia

Unless a force acts on an object, the object's motion doesn't change. If it is at rest, it stays at rest. If it is moving, it carries on moving at a steady speed in a straight line. We call this observed property of all objects **inertia.** We cannot explain inertia – its just the way all objects behave!

An example: forces in cycling

Here is an example showing how forces are involved in steady motion and changes in motion.

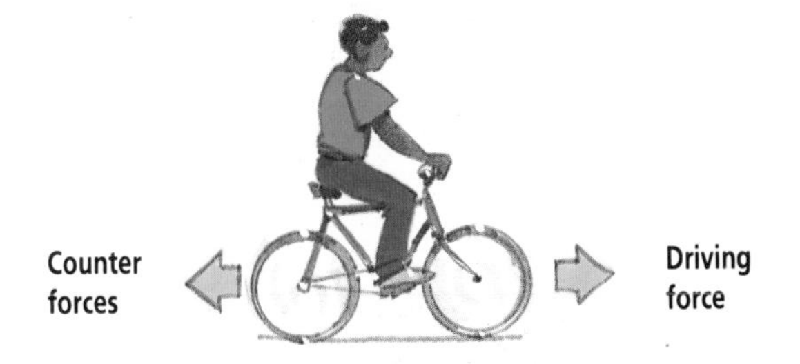

If you are cycling along a road at a steady speed, the forces on you are balanced. The driving force is exactly equal to the counter forces (friction and air resistance). Air resistance is a special case of friction.

If you want to speed up, you have to exert a larger driving force. This is now bigger than the counter forces, so you speed up.

On the other hand, if you want to slow down, you pedal less hard. The counter forces are now bigger than the driving force. The total force is backwards — against the motion. So you slow down.

3. Terminal velocity

Forces are balanced when you are stopped, and forces are balanced when you are riding along at a steady speed. What happens in between? How do you ever get moving in the first place?

When you start to pedal a bicycle you provide a driving force. At this stage the counter forces are small. As you speed up, the counter forces get bigger. You stop speeding up when the counter forces are the same size as the driving force. The speed you are travelling at then is called the **terminal velocity.**

When you start off, the driving force is bigger than the counter forces, so you speed up.

As your speed increases, the counter forces increase. You continue to speed up, but not as quickly as before.

Eventually you reach a speed where the counter forces are exactly equal to the driving force. Now your speed stops increasing. But you don't stop — you keep on travelling at this speed.

4. What is acceleration?

If no force acts on an object (or if the forces are balanced and add to zero), the object's motion stays the same. But if a force does act on an object, its motion changes. A steady force in the direction the object is moving makes it speed up or **accelerate**. A steady force in the opposite direction to its motion makes it slow down – a negative acceleration.

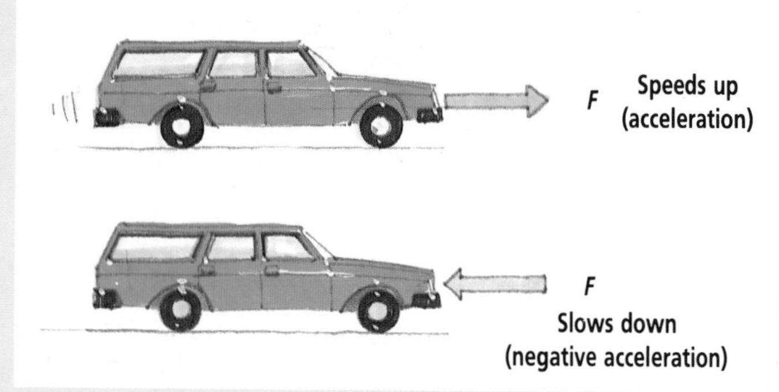

Acceleration is defined by this equation:

$$\textbf{acceleration} = \frac{\textbf{change of velocity}}{\textbf{time taken}}$$

Here are some examples of this equation in action.

Example 1: A car can accelerate from 0 to 60 m.p.h. in 8 seconds. What is its acceleration?

$$\textbf{acceleration} = \frac{\textbf{60 m.p.h.}}{\textbf{8 s}} = \textbf{7.5 m.p.h. per second}$$

This means that the car's speed increases by 7.5 m.p.h. every second.

Example 2: An ice skater can speed up from rest to 12 m/s in 8 s. What is her acceleration?

$$\textbf{acceleration} = \frac{\textbf{12 m/s}}{\textbf{8 s}} = \textbf{1.5 m/s per second}$$

This means that her speed increases by 1.5 m/s every second.

Example 3: A car is travelling at 25 m/s when the driver brakes. The car stops in 5 s. What is the acceleration?

$$\textbf{acceleration} = \frac{\textbf{– 25 m/s}}{\textbf{5 s}} = \textbf{– 5 m/s per second}$$

The negative sign means that the car's speed is getting *less* by 5 m/s every second.

5. How are force, mass and acceleration connected?

If a force acts on an object, the object accelerates. But how big is the acceleration? It depends on the size of the force and the mass of the object. We can investigate this in the laboratory.

(a) Force and acceleration

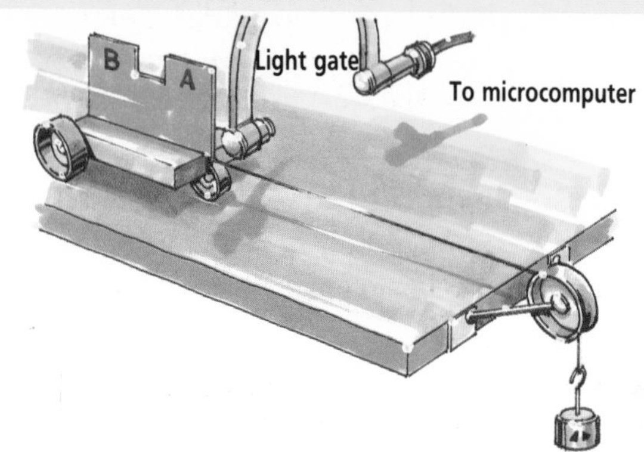

The mass on the end of the thread exerts a force on the trolley. We can increase the force by adding more masses. We measure the acceleration by fixing a double card to the trolley. Part A cuts through a light gate and interrupts the beam. Then part B interrupts it again. The computer measures the times taken for A and B to cut through the beam. If it knows the width of A and B, it can work out the two speeds – one when A goes through, the other when B goes through. It also measures the time interval between these. The computer can then work out the acceleration of the trolley.

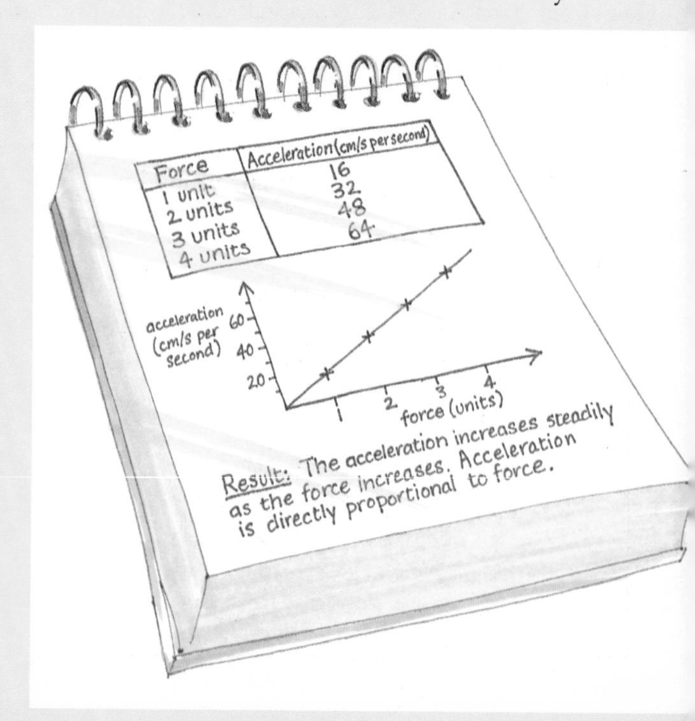

(b) Mass and acceleration

This time we keep the pulling force constant, but change the mass on the moving object by stacking more trolleys on top.

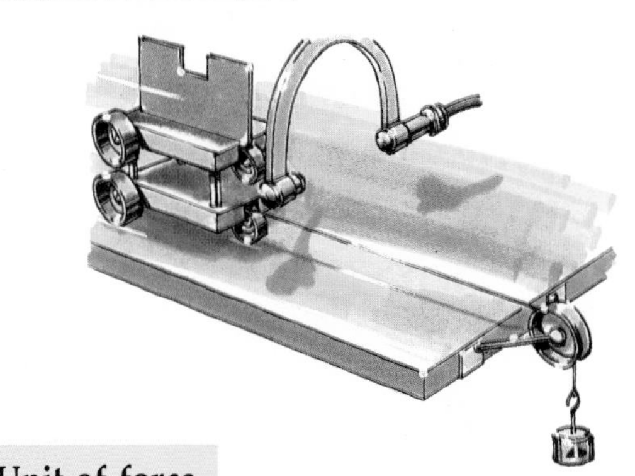

Mass	Acceleration (cm/s per second)
1 trolley	36
2 trolleys	18
3 trolleys	12

Result: The acceleration gets smaller as the mass gets bigger. If we double the mass, the acceleration is roughly half of its previous value, and so on. So we have to double the force to give it the same acceleration as the single trolley. With three times the mass, the force has to be three times as large to give the same acceleration (Another way of putting this is to say that acceleration is inversely proportional to mass).

Unit of force

A force is something which makes a mass accelerate. So we can define the unit of force, the **newton** (N) as follows:

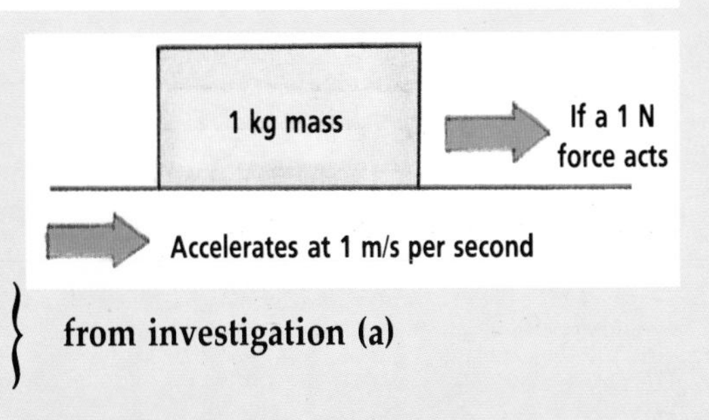

a 1 N force is needed to accelerate 1 kg at 1 m/s per second

So

a 2 N force is needed to accelerate 1 kg at 2 m/s per second
a 3 N force is needed to accelerate 1 kg at 3 m/s per second } **from investigation (a)**

and

a 2 N force is needed to accelerate 2 kg at 1 m/s per second
a 3 N force is needed to accelerate 3 kg at 1 m/s per second } **from investigation (b)**

or

a 4 N force is needed to accelerate 2 kg at 2 m/s per second
a 6 N force is needed to accelerate 2 kg at 3 m/s per second } **from both investigations together**

In all cases, **force = mass × acceleration**

$$F = m\,a$$

This is called **Newton's second law of motion.**

Taking it further:

There is another very useful way of stating Newton's second law. Imagine a trolley changing its speed from $v_{initial}$ to v_{final} in t seconds.

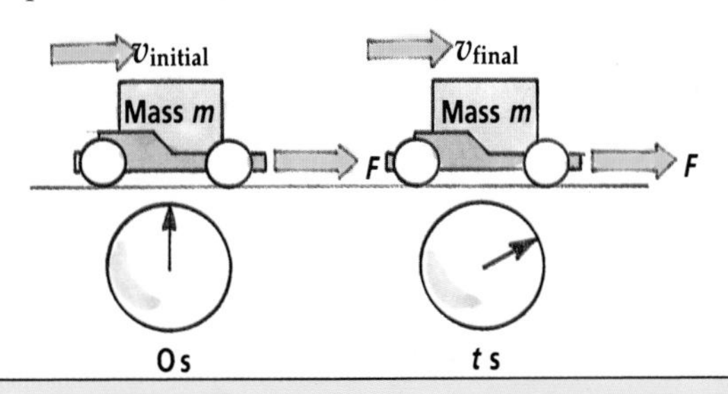

Its acceleration is:

$$a = \frac{(v_{final} - v_{initial})}{t}$$

If we substitute this into the equation for Newton's second law above:

$$F = \frac{m\,(v_{final} - v_{initial})}{t}$$

or

$$Ft = mv_{final} - mv_{initial}$$

We will see in the next section how to use this equation.

6. What happens when you stop suddenly?

Sometimes things change speed very suddenly, for example when a car crashes. A very useful equation for helping us work out what happens then is:

$$Ft = mv_{\text{final}} - mv_{\text{initial}}$$

F is the force which causes the change of speed and t is the time it takes for the change to happen.

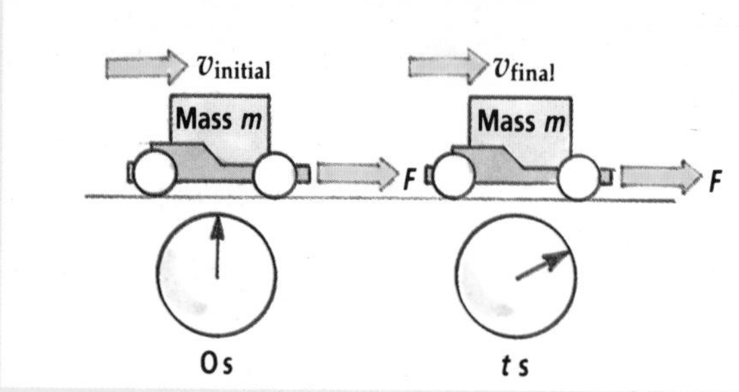

Now imagine a car of mass m travelling along at speed v_{initial} when it is suddenly brought to a stop.

t is the actual duration of the impact – from the moment the car first makes contact with the wall until it stops moving. The equation now becomes simpler, because the final speed v_{final} is zero:

$$Ft = -mv_{\text{initial}}$$

The car's mv_{initial} depends on its mass and its speed before the crash. If we know these we can work it out. This equation tells us that F multiplied by t must come to the same number. So if the time taken for the car to stop is small, the force will be large. But if the time is longer, the force will be smaller.

Many modern cars have 'crumple zones'. The front of the car crumples up in a collision, which increases the time the car takes to stop. Other safety features of cars work in the same way. They lengthen the time it takes for a moving object to stop, so that the force is smaller. Seatbelts, crash barriers and safety helmets all make use of this idea. Pages 60–61 and 64 tell you about this in more detail.

7. What happens in a collision?

How hard is it to stop a moving object? The amount of mv it has tells you. If it has a lot of mv, it is hard to stop. If it has only a little mv, it is easier to stop. The quantity mv is given a special name: **momentum.**

Work out the momentum of each of the vehicles in these photos. Which would be the easiest to stop? Which needs the most powerful engine to get it going at this speed?

Momentum plays an important role when two objects collide. The total momentum after the collision is the same as the total momentum before.

Before | After
120 mm/s | 60 mm/s
120 mm/s | 80 mm/s
120 mm/s | 90 mm/s

Mass (kg)	Velocity (mm/s)	mv (kg mm/s)
1	120	120
2	120	240
3	120	360

Mass (kg)	Velocity (mm/s)	mv (kg mm/s)
2	60	120
3	80	240
4	90	360

Explosions, where two objects spring apart, are a special type of collision. The total momentum at the beginning is zero – everything is stationary. Afterwards, the two objects have equal and opposite momenta. Direction is important for momentum! Equal and opposite momenta add to zero.

A	B	$(mv)_A$ (Kg mm/s)	$(mv)_B$ (Kg mm/s)
200 mm/s	200 mm/s	−200	200
230 mm/s	115 mm/s	−230	230
246 mm/s	82 mm/s	−246	246

Taking it further: momentum and energy

All moving things have **kinetic energy** and **momentum**. But momentum is *not* the same as kinetic energy. Think about the explosion experiment. Before the explosion, there was no kinetic energy – everything was standing still. There was also no momentum. After the explosion, the two momenta were equal and opposite. One has a negative sign and they add to zero. But energy does not have direction. Both trolleys have kinetic energy and this does *not* add to zero.

The explosion is an energy transfer. The stored elastic potential energy in the spring is transferred to kinetic energy in the moving trolleys.

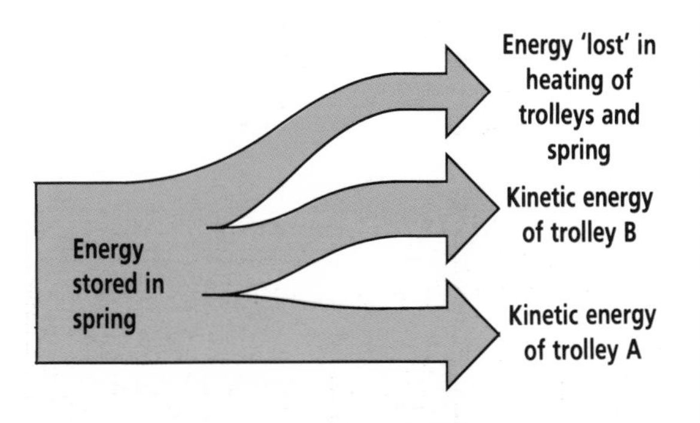

8. Pressure

When a force pushes on an object, the force exerts a **pressure** on the surface of the object. The size of this pressure depends on how big the force is and the area it pushes on.

We define pressure using the equation:

$$\textbf{pressure} = \frac{\textbf{force}}{\textbf{area}}$$

The units of pressure are newtons per metre squared (N/m^2), or newtons per centimetre squared (N/cm^2). 1 N/m^2 is also called 1 **pascal** (Pa).

Here are two examples, one showing a high pressure and the other a small pressure.

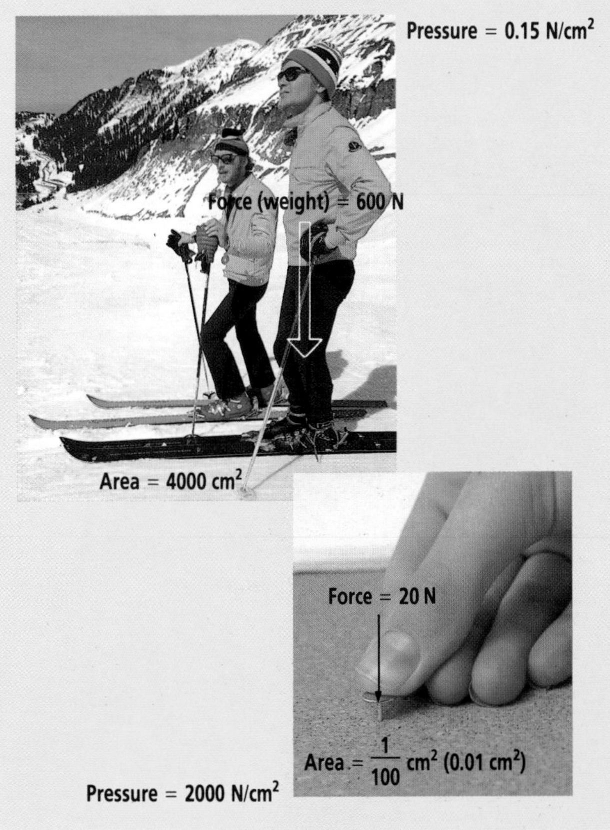

It is the pressure, not the force, which determines how far the pushing object sinks into the other surface. Of course, it also depends on how hard the other surface is. The force exerted by the drawing pin on the wallboard is smaller than the downward force of the skier on the snow. But the pressure of the pin on the board is much bigger than the skier's pressure on the snow. So the pin goes into the board, but the skier does not sink into the snow.

Things to do

Moving On

Things to try out

1 Here are some tricks you can practise and then try out on your friends. They all have a scientific explanation!

(a) Get a medium-sized hardback book and a reel of thread. Tie a length of thread to hold the book. Tie a second length of thread to hang beneath the book.

Now hold the book by the top thread. (You might need to tie a loop in the thread, so that it does not slip out of your fingers. Wearing a glove will also help if the book is heavy.)

Pull gently on the lower thread. Keep pulling . . . which thread breaks first?

Now try this again – but this time pull the lower thread with a sudden jerk. Which thread breaks this time?

(b) Make a pile of pieces from a draughts set. Place this on a smooth tabletop. Take a ruler and quickly sweep it along the table. Can you knock out the bottom draughts piece but leave the others standing?

Try to write a scientific explanation of how these tricks work. The scientific idea of inertia is important in each case.

2 To try this out, you need a bicycle and a small hill on a quiet road near your home. You have to start at the same point each time, somewhere up the hill. So mark a spot on the ground with chalk. Now freewheel down the hill (no pedalling!) and continue freewheeling as far as you can along the level before you stop. Mark the furthest point you reach with chalk.

Now go back up the hill to the same starting place. This time let some air out of your tyres until they are only about half pumped up. Try freewheeling down the hill again.

Do you go as far this time? Give a reason for the difference, using science ideas that you have learnt in this unit.

(P.S. Don't forget to pump your tyres up again!)

3 Find a fairly heavy toy car or lorry and an elastic band. Pull the vehicle along with the elastic band and see how long the band gets. You should notice that it is longer when the vehicle is starting than when it is moving steadily.

Try to explain this using the scientific ideas you have learnt in this unit.

Points to discuss

4 In a road accident a car swerves to avoid a child and runs into a wall. The front of the car is completely crumpled in, but the driver is able to open her door and step out uninjured.

A man who saw the accident comments, 'They don't make cars like they used to. Look at that! The front of the car is completely ruined.'

Is the man right in thinking that the car has crushed because it is made of cheaper materials? How would you explain to him why the car is designed to crush easily?

5 Explain each of the following common observations using science ideas in your explanations. Use a diagram if necessary.

(a) If you are standing on a bus when it starts moving, you tend to fall backwards. If you are standing on a moving bus when it suddenly stops, you are thrown forwards.

(b) If you have to brake suddenly on your bicycle, you almost fly off over the handlebars. The back wheel may even lift off the ground.

(c) (Harder) If you are in a car which goes round a corner quickly, you feel as if you are being pushed outwards, towards one side of the car.

6 Some of the worst road accidents happen when a moving car hits a solid fixed object, like the pillar of a motorway bridge. Road engineers have been developing a crushable crash barrier to reduce this danger. The picture shows this barrier being tested.

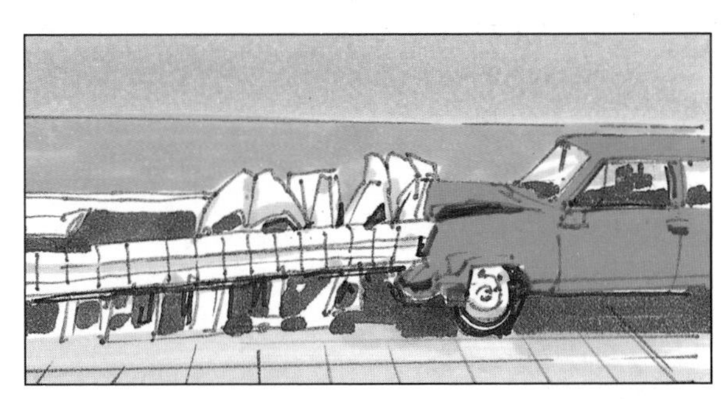

Make a plan drawing of a motorway bridge, showing where you would place a crushable crash barrier.

Explain, using science ideas from this unit, how this sort of barrier works, and how it reduces injury and damage in a crash.

Things to write about

7 Cut out a car advertisement from a newspaper (there are usually lots in the Sunday supplements). Choose one which emphasises the safety aspects of the car.

Make this cutting the centre-piece of a poster. Around the cutting, add some extra information to explain the car's safety features.

8 The pictures show two cars after crash testing. The front end of the car in the top picture was not designed as a crumple zone. The lower car did have a crumple zone design.

Make a list of all the differences you can see between the damage done to the two cars.

Which car would be safest to drive?

Explain how the crumple zone works, using science ideas you have learnt from this unit.

Questions to answer

9

What force is needed to make the discus accelerate at:

(a) 2 m/s^2 **(b)** 0.5 m/s^2 **(c)** 10 m/s^2?

10 Put the vehicles in the photos below into order, from the largest acceleration to the smallest acceleration.

Boeing 747:
Mass 400 000 kg
Engine force 800 000 N

Ford Fiesta XR2i
Mass 1300 kg
Engine force 6500 N

BMW 1000
Mass 300 kg
Engine force 3000 N

Car ferry
Mass 2 000 000 kg
Engine force 200 000 N

Cyclist
Mass 90 kg
Driving force 135 N

11 A loaded supermarket trolley has a mass of 20 kg. It is rolling along at a speed of 3 m/s, when it bumps into an empty trolley of mass 10 kg.

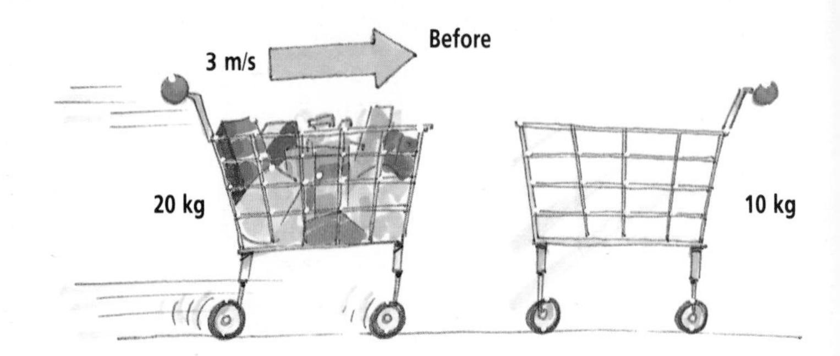

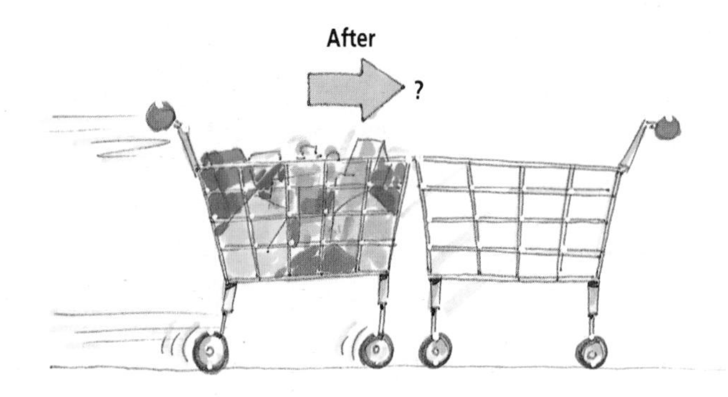

(a) What is the momentum of the first trolley?

(b) If the two trolleys stick together after they crash, what speed will they be moving at?

12 What determines how deep a mark a table makes on a carpet: its weight or the pressure its feet exert?

Which of these tables would make the deepest mark on a carpet?

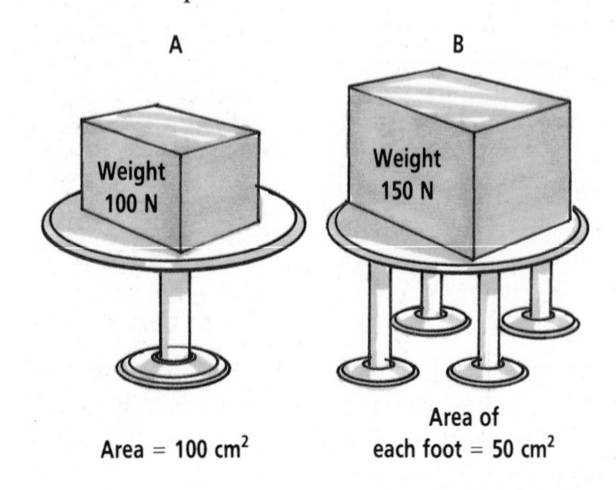

Introducing
FOOD FOR THOUGHT

You probably take it for granted that there will be something for tea tonight. But we see television reports that some people in the world are starving, and many organizations like Band Aid and Oxfam work hard to raise money to help them. There are 5 billion people in the world, and we *do* have enough food to feed them all.

1. Suggest three reasons why some areas of the world have a food shortage while others have plenty of food.
2. In 1900 there were about 2 billion people in the world and there wasn't enough food to go around. Now there is enough for everybody. Make a list of reasons why we can now produce so much more food. The pictures on this page will help you.
3. Scientists have developed ways of producing enough food. How can we make sure the people who need it will get it? Draw a cartoon story explaining who can help and how.

IN THIS CHAPTER YOU WILL FIND OUT

- how we can increase the amount of crops grown on a piece of land (the crop yield)
- how nitrogen-containing fertilizers work and how they are produced
- how foods are processed and preserved.

Looking at

Increasing Food Production

Where does our food come from?
If you think all food comes from supermarkets, think again! To make sure everyone has enough food, we have to look at where the food comes from and how we can produce more of it.

Photosynthesis
Green plants are the original source of all our food. They make their own food by a process called **photosynthesis.** Their leaves are green because they contain a chemical called **chlorophyll.** It can capture light energy and use it to make carbohydrate (sugars). This reaction summarizes photosynthesis.

carbon dioxide + water $\xrightarrow{\text{sunlight}}$ carbohydrate + oxygen

1 If you eat nothing but meat (not a healthy diet!) does this mean your food doesn't come from green plants?

Plants use the food they make to grow. We produce crop plants to feed ourselves and farm animals. The more carbohydrate a plant produces the more it grows and the better the crop yield. If we can increase the amount of photosynthesis then we can increase the amount of carbohydrate, improve the crop yield and produce more food. How can we do this? Follow the arrows to find out.

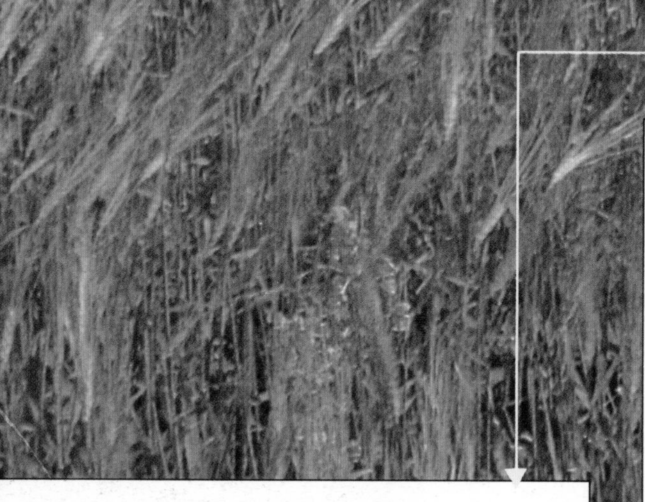

CHOOSE THE BEST PLANT

All living things show variation. Some individual plants in a crop give higher yields than others. By selecting those plants which give the highest yields and breeding them together, new high-yielding varieties can be produced to suit particular conditions. This is one of the main ways in which food crop production has been improved.

2 Imagine you were given the task of breeding a variety of corn to be grown in Ethiopia. What conditions would it have to be able to survive? Where might you start looking for varieties?

GIVE IT MORE LIGHT

Photosynthesis improves when there is plenty of light. It is not practical to place floodlights over fields but artificial lights can be used in greenhouses.

3 These lights are usually controlled to come on in the morning and evening. Why is this?

Another way of increasing photosynthesis is to make sure that most of the light is captured by the plant's chlorophyl. Plants with upright leaves are better at capturing light than plants with horizontal leaves. Plants with leaves that point upwards can be grown close together as the leaves do not overlap. Close-planted crops make better use of the available land and the light.

4 Sketch diagrams to show that: (a) plants with upright leaves can collect more light than plants with horizontal leaves (b) plants with upright leaves can be planted closer together than plants with horizontal leaves.

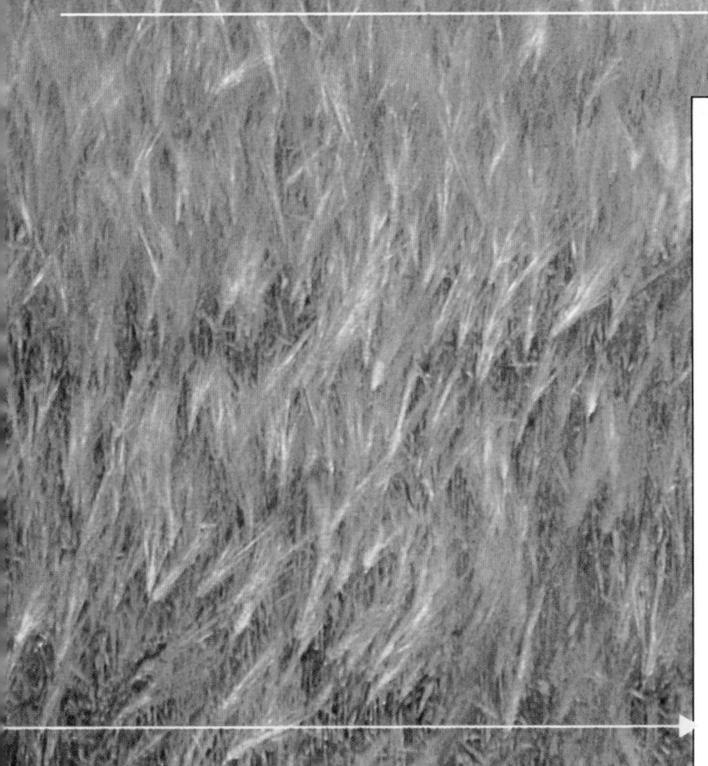

GIVE IT MORE WATER

Crops need plenty of water for photosynthesis. In hot dry climates water can be provided by **irrigation.** About 70% of the world's water supply is used for irrigation.

7 List the advantages and disadvantages of each method of irrigation described below.

- **Irrigation canals** bring streams of slow-moving water close to crop fields. They are easy to construct but provide a good place for snails to live. Snails can carry human diseases.
- **Centre-pivot systems** can provide large amounts of water. Too much water can dissolve salts from the ground and leave them in the topsoil when it evaporates. Plants will not grow when the topsoil contains a high concentration of salts.
- **Trickle-feed systems** allow water to drip from pipes placed close to plants. Little water is wasted through evaporation but systems are very expensive to install.

KEEP IT WARM

Increasing the temperature increases the rate of chemical reactions, including photosynthesis. Greenhouses warm up by the heat of the Sun. They are used to grow food plants such as tomatoes. A greenhouse-grown tomato plant will give a greater yield than one grown outdoors. If greenhouses are heated artificially certain food plants can be available for a longer season. Heating can be provided by an electric heater or oil or gas burner inside the greenhouse, or hot water pipes like a central heating system.

5 How would you choose to heat (a) a small garden greenhouse (b) a large commercial greenhouse?

GIVE IT MORE CARBON DIOXIDE

Increasing the amount of carbon dioxide around plants can increase the amount of photosynthesis. This is not easy to do in a field but it can be done in a greenhouse.

6 Draw a diagram showing how greenhouse crops could be kept warm and have lots of carbon dioxide.

Reaching the limit

Water, carbon dioxide, light and temperature conditions *all* have to be right for the best crop yield. There is no point giving more water if the low temperature is limiting growth! Another factor is the soil. Plants need certain chemicals from it to be healthy.

8 What can we do about poor soils?

Looking at

Food Emulsions

Dear Editor,

I approve of the recent campaign in school for healthier eating and so I was horrified when I read the label on the Sunbeam Margarine which I had with my baked potato yesterday. The margarine sold in the school dining room contains an additive called an emulsifier. Why do we have to put up with food with emulsifiers in them? What are they? I thought emulsions were something to do with paint?

Yours

This letter has been published in the school newspaper. You have been asked by the editor to write an article in reply. The article should explain what emulsifiers and emulsions are. It should also give reasons whether or not people ought to be concerned about their use. On this page and the next there is some information that will help in writing the article.

What is an emulsifier?

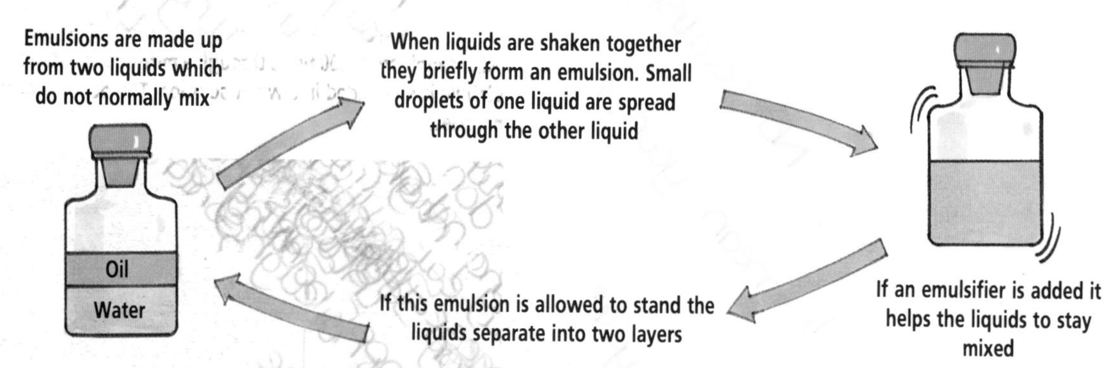

Emulsifiers are usually long molecules with different groups of atoms at either end.

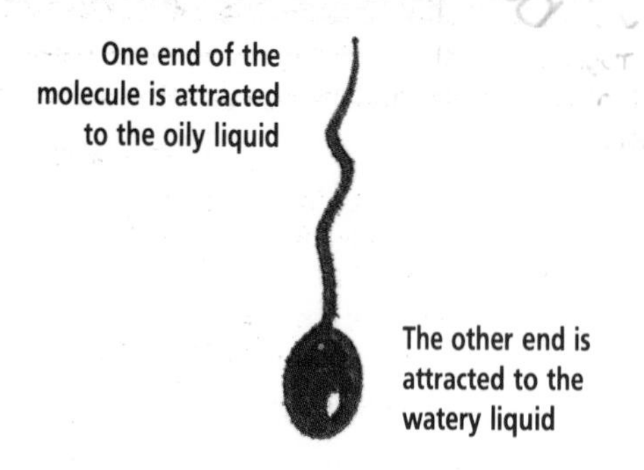

How emulsifiers work

The diagram shows how the emulsifier keeps the two liquids A (oily) and B (watery) mixed.

The oil is broken up into small droplets. Each droplet is surrounded by molecules of the emulsifier.

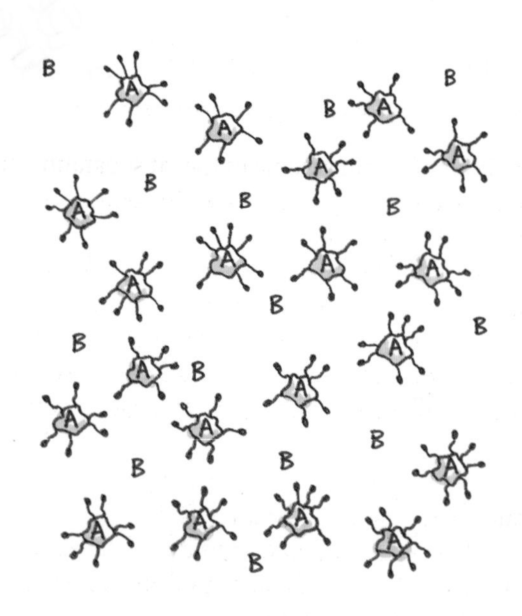

Salad dressings

Some people like to put a mixture of vinegar and oil on their salad. Vinegar and oil do not mix but if you shake them together they form an emulsion. But this emulsion separates into two layers.

An emulsifier is needed to make it last longer. The emulsifier in 'French dressing' is mustard.

Milk and butter

Milk is a natural emulsion. It contains a watery part and an oily part. The oil is known as butter fat. This oil in water emulsion is held together by the protein molecules contained in milk. These protein emulsifiers keep the fat dispersed as small droplets.

If milk is allowed to stand a layer of cream floats to the top.

The layer of cream is some of the fat separating out from the emulsion. If you shake the bottle you can re-make the emulsion.

The cream can be skimmed off to make a mixture with a higher proportion of butter fat. If this mixture is shaken (churned) it eventually changes from an oil in water emulsion to a water in oil emulsion. This water in oil emulsion is butter.

Many foods you eat contain a mixture of fat and water. They are more pleasant if the fat and water do not separate. If you examine food labels you will see which foods contain emulsifiers to stop the fat and water separating.

Egg white contains the protein lecithin, which is an emulsifier. Eggs are added to oil and vinegar to make mayonnaise.

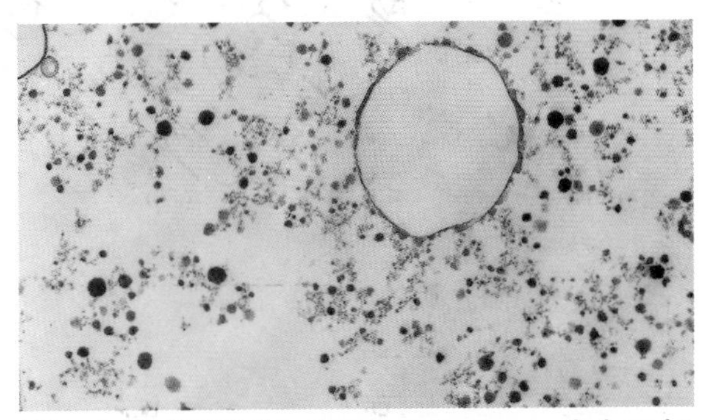

Milk magnified 25 000 times through a microscope — the large fat globule is suspended in a water solution. The dots are protein particles.

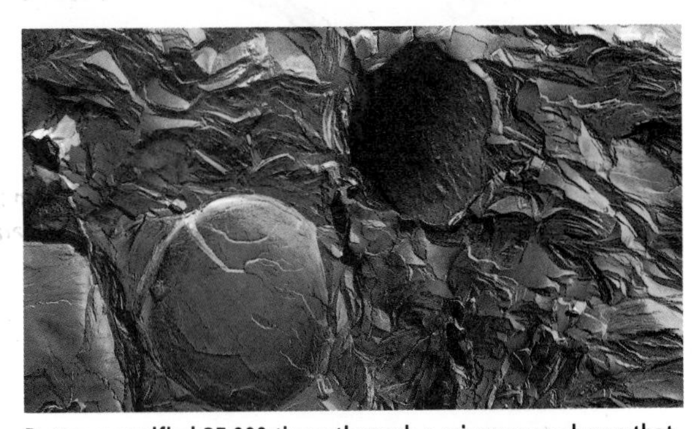

Butter magnified 35 000 times through a microscope shows that droplets of water are dispersed through the fat — a water in oil emulsion.

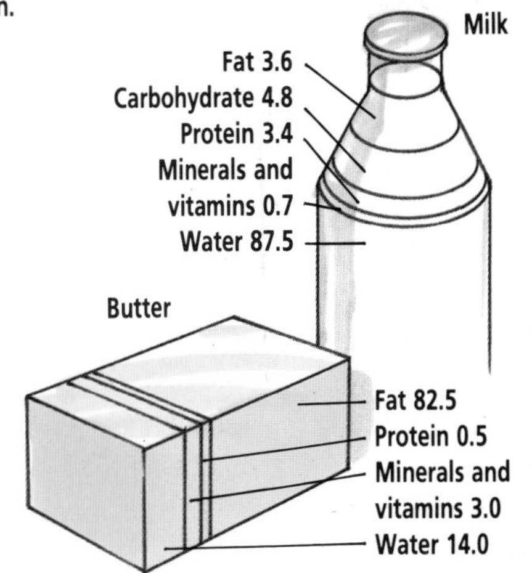

In brief

Food for Thought

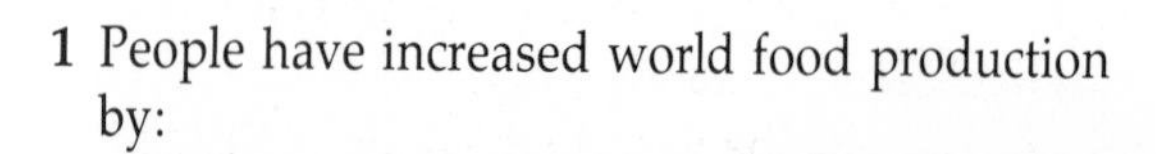

1 People have increased world food production by:

2 Growing and harvesting crops removes nutrients from the soil which are essential for healthy growth.

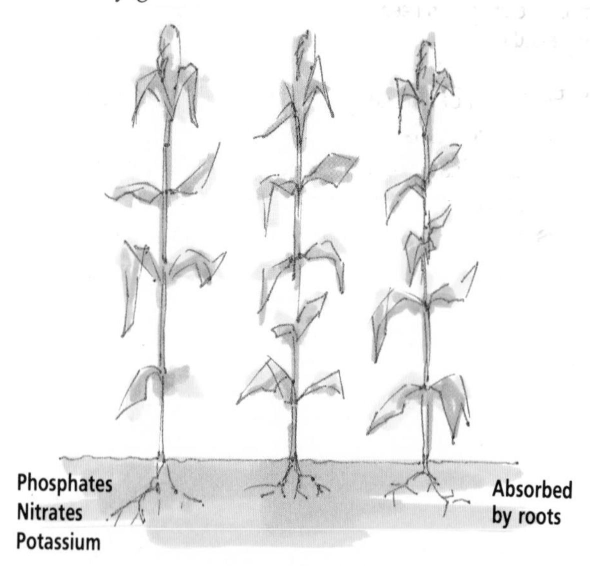

Nutrient-deficient soil can be improved by adding chemical fertilizers. Most chemical fertilizers contain compounds of nitrogen, potassium and phosphorus.

3 Compounds of nitrogen are particularly good for encouraging plant growth. Although there is plenty of nitrogen gas in the air, most crop plants cannot use it directly. But the nitrogen can be combined with hydrogen to form ammonia. The industrial process developed by Fritz Haber is used to do this.

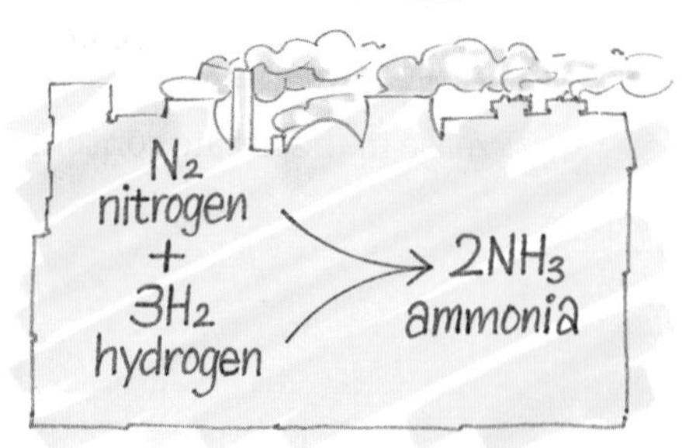

4 Ammonia can be used as a fertilizer, but it is a **base** – it dissolves in water to form an alkaline solution. This means it would change the pH of the soil. It is also smelly and difficult to handle.

This problem is overcome by reacting ammonia with acids to form **salts.** These are solids and almost neutral. This reaction is called a **neutralisation.**

5 Fertilizer manufacturers need to know how much of each chemical to react together to produce the required quantity of the fertilizer. This can be calculated from the balanced equation for the reaction and the relative molecular masses of the substances involved.

$$2NH_3 + H_2SO_4 \rightarrow (NH_4)_2SO_4$$

RMM 34   98   132

34 tonnes of ammonia and 98 tonnes of sulphuric acid give 132 tonnes of ammonium sulphate.

Relative molecular masses are calculated from the relative atomic masses of the elements in the molecule.

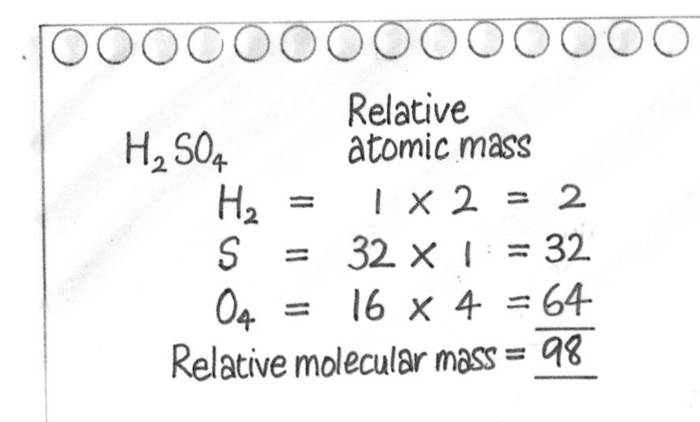

6 Because salts used as fertilizers are soluble, they may be washed or **leached** out of soil and reach streams and rivers, causing pollution. For example, an increase in the nitrate concentration of a river can result in unwanted growths of algae and a lower quality of drinking water.

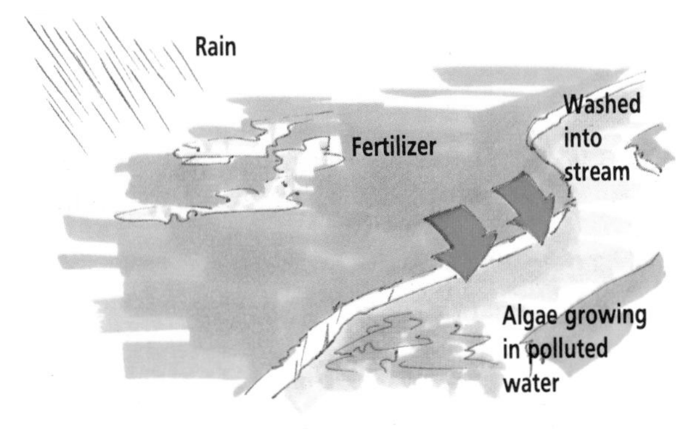

7 To keep our stored food fit to eat we need to prevent the growth of microbes and keep out animal pests. Many animal pests are controlled by chemical pesticides, but non-chemical means can also be used. For example, instead of spraying aphids (greenfly), ladybirds can be released on the crop to eat the aphids. Some foods have preservatives added to them to make them keep longer.

Locusts can devastate crops. Chemical sprays are used to kill them.

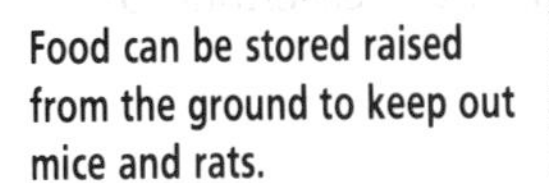

Food can be stored raised from the ground to keep out mice and rats.

8 Food can be processed to make it more useful or attractive. This can be done by

- chemical reactions, such as adding hydrogen to vegetable oil to make margarine
- biological reactions, such as using the enzyme pectinase to extract juice from apples
- food additives which improve the flavour or appearance of the food.

Recently people have found that some food additives can be harmful. However, foods go 'off' much more quickly without preservatives.

Thinking about

Food for Thought

1. How can we improve the growth of crops?

Plants need certain elements from the soil to be healthy and grow well. Soil which does not contain these essential nutrients is infertile. Fertilizers are chemicals which farmers add to the soil to provide the right nutrients. Sometimes natural fertilizers like manure or compost are used instead of chemical fertilizers. Growing crops and harvesting them removes

- nitrogen (N)
- phosphorus (P)
- potassium (K)

from soil, so these elements have to be replaced.

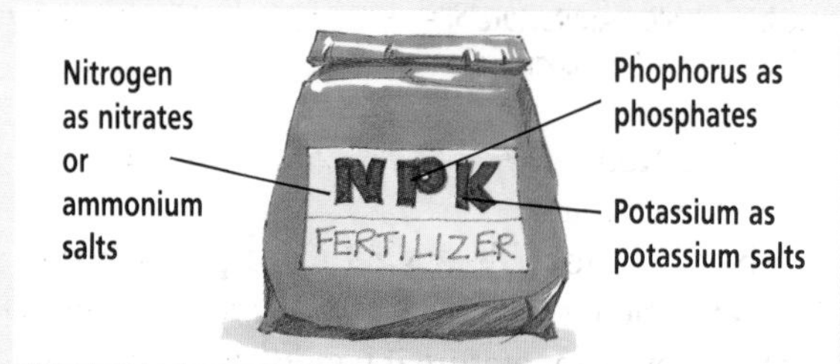

Most manufactured fertilizers contain these three elements.

Chemicals are also used to protect healthy plants from damage. Pesticides are substances which kill pests and herbicides are substances which kill weeds.

Pests spoil or eat up to one-third of the world's food while it is being grown, harvested or stored. These Colorado beetles are eating potato leaves.

In some regions pests can waste up to 40% of crops. Pesticides can help to reduce this loss. However pesticides must be chosen with care. They must break down into non-harmful chemicals. They must not destroy non-harmful animals.

2. How are fertilizers manufactured?

The raw materials for making fertilizers are either dug out of the ground or extracted from the air.

- Phosphorus comes mainly from rock phosphate – a mineral containing calcium phosphate.
- Potassium comes from minerals such as sylvinite which contains potassium chloride.
- Nitrogen, as you might expect, comes from the air.

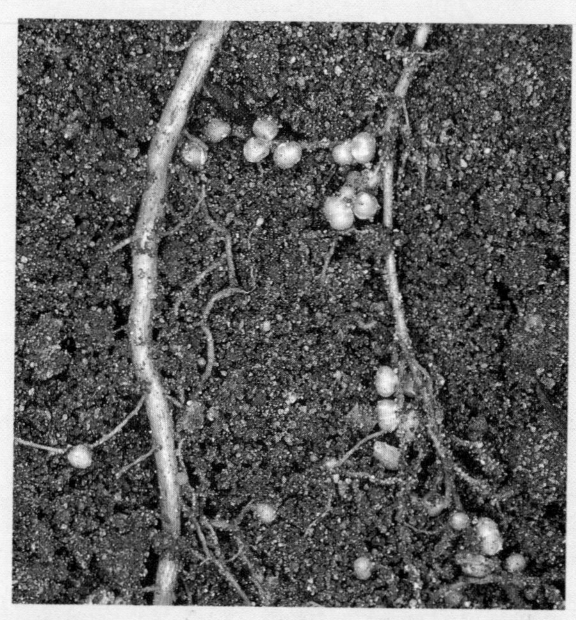

Nitrogen is converted into a usable form by bacteria in nodules on the roots of some plants such as peas and beans.

Lightning converts some nitrogen and oxygen from the air into oxides of nitrogen and then nitrates.

Bacteria and lightning cannot replace all the nitrogen in soil that crops use. So we need to manufacture fertilizers from nitrogen.

People realized they needed to 'fix' nitrogen from the air (convert it to a form plants can use) in the early part of this century. A German chemist, Fritz Haber, invented a method of converting nitrogen and hydrogen into ammonia (NH_3).

Fritz Haber used small-scale equipment to develop the process.

The large-scale manufacture of ammonia in this fertiliser factory is still based on the Haber process.

By 1913 an engineer, Carl Bosch, had scaled up Haber's process so that ammonia could be produced on an industrial scale. The process is still known as the Haber process.

In the modern version of the process a mixture of hydrogen and nitrogen is passed over a **catalyst** at 400°C and 200 atmospheres pressure (this is 200 times greater than normal atmospheric pressure). A catalyst speeds up a chemical reaction without being used up itself. Iron is the catalyst for the Haber process. The reaction is **reversible** – this is shown by the sign $\rightleftharpoons$

$$N_2 + 3H_2 \rightleftharpoons 2NH_3$$

A reversible reaction can be encouraged to go in either direction by changing the conditions.

If the reaction moves to the right the yield of ammonia increases.

The yield would be higher if the pressure was high and the temperature low. But the problem is that at a low temperature it would take a long time for the ammonia to form. A reasonably high temperature together with a catalyst produce the ammonia more quickly.

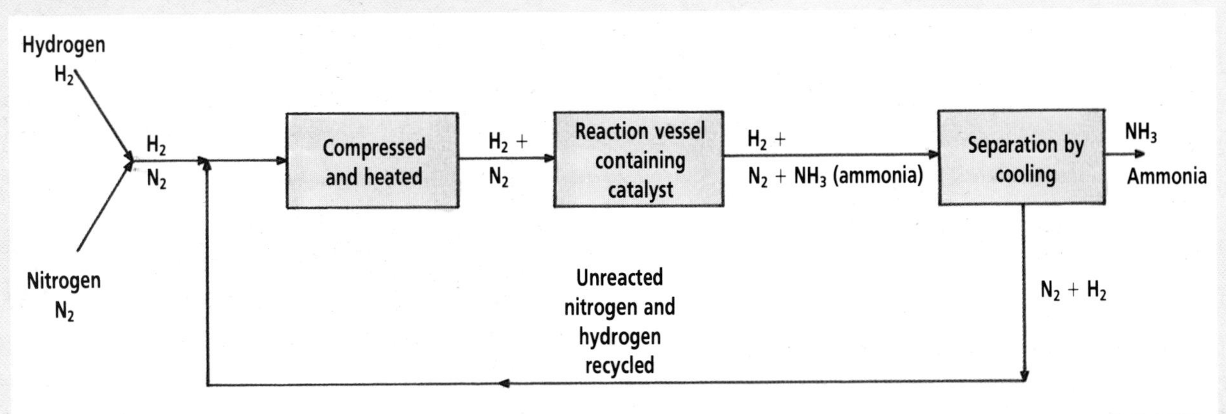

The ammonia is removed by cooling it to a liquid. The unreacted hydrogen and nitrogen are re-circulated so they are not wasted.

3. How can we make ammonia into a more useful fertilizer?

Pure ammonia is a very unpleasant smelly gas. High concentrations of it in the air would be extremely dangerous.

It dissolves easily in water, but even the solution smells very strongly of ammonia. The solution is alkaline so it would change the pH of soil if it was used as a fertilizer.

To make a more useful fertilizer we need to convert ammonia into a solid which does not smell much, is not washed away as quickly as ammonia and does not have much effect on the pH of soil.

The substance formed when any acid reacts with a base is known as a salt. Ammonium sulphate and ammonium nitrate are both salts. They are used as fertilizers. They are solids and so are easier to handle than ammonia. They do not upset the pH of soil. Reactions between acids and bases are called **neutralisations.**

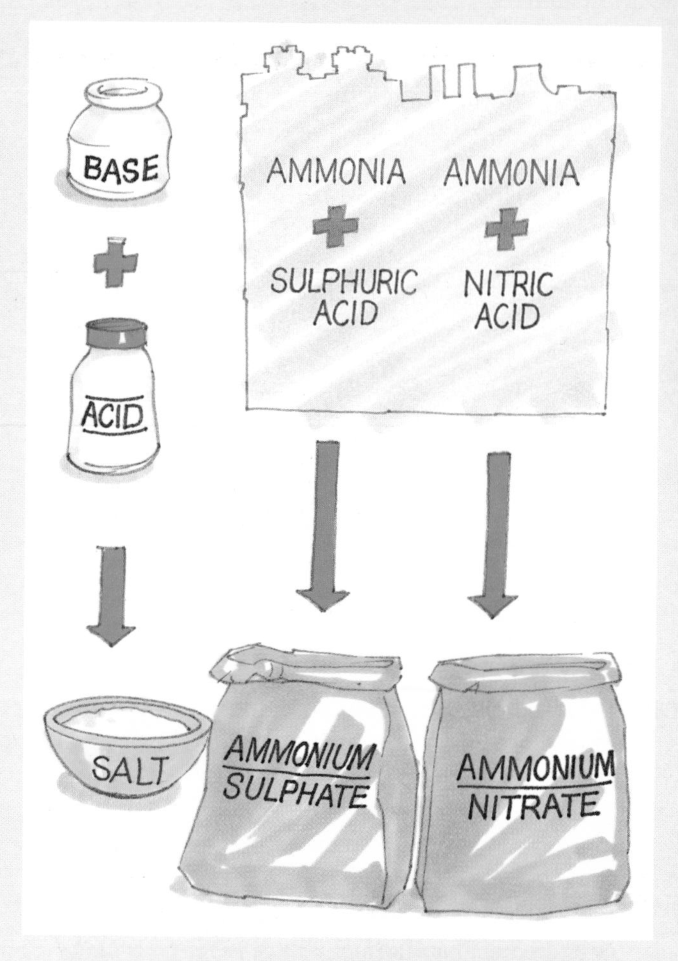

Taking it further

All acids contain hydrogen. **Salts** are formed when these hydrogens are replaced by metals. The name of the salt formed depends on the acid used and the metal which replaces the hydrogen.

It is easy to work out the name of a salt formed from an acid and a base. The table gives the names and formulas of salts formed from sodium hydroxide with three different acids.

Sodium hydroxide is a soluble base and so it is called an **alkali**.

Acid	Formula	Salt	Formula
hydrochloric acid	HCl	sodium chloride	$NaCl$
nitric acid	HNO_3	sodium nitrate	$NaNO_3$
sulphuric acid	H_2SO_4	sodium sulphate	Na_2SO_4

When these acids react with a solution of ammonia, the ammonium group replaces the hydrogen. The salts formed are

- ammonium chloride, NH_4Cl
- ammonium nitrate, NH_4NO_3
- ammonium sulphate, $(NH_4)_2SO_4$.

Sodium chloride is one of many salts. It is called common salt, or often just 'salt' or 'table salt'.

4. How much ammonia is needed to make fertilizers?

Memo: FROM JANE SMITH (MANAGER)
This month we need to produce 264 tonnes of ammonium sulphate. Please let me know the amounts of ammonia and sulphuric acid we will need.

Memo: FROM LIAM O'GRADY (PRODUCTION)
The balanced equation for the reaction is:

$$\underset{\text{ammonia}}{2NH_3} + \underset{\text{sulphuric acid}}{H_2SO_4} \rightarrow \underset{\text{ammonium sulphate}}{(NH_4)_2SO_4}$$

The relative atomic masses of these elements are:

H = 1, N = 14, S = 32, O = 16

The relative molecular masses are:

NH_3	=	14 + (3 x 1)	=	17
H_2SO_4	=	(2 x 1) + 32 + (4 x 16)	=	98
$(NH_4)_2SO_4$	=	2(14 + 4) + 32 + (4 x 16)	=	132

The equation for the reaction shows that 2 molecules of ammonia react with 1 molecule of sulphuric acid to form 1 molecule of ammonium sulphate.

So (2 x 17) = 34 tonnes of ammonia react with 98 tonnes of sulphuric acid to form 132 tonnes of ammonium sulphate.

So for 132 tonnes of ammonium sulphate, you need 34 tonnes of ammonia and 98 tonnes of sulphuric acid.

For 264 tonnes of ammonium sulphate you need (2 x 34) = 68 tonnes of ammonia and (2 x 98) = 196 tonnes of sulphuric acid.

5. How is food processed chemically?

Oil is extracted from sunflower seeds. It is good for frying but too runny to spread on bread.

The oil is processed to make a soft solid which can be spread on bread.

Oil molecules are very long and have double bonds in them. Molecules with double bonds are called **unsaturated** – hydrogen can add on across the double bonds. Molecules with lots of double bonds are called **polyunsaturated**.

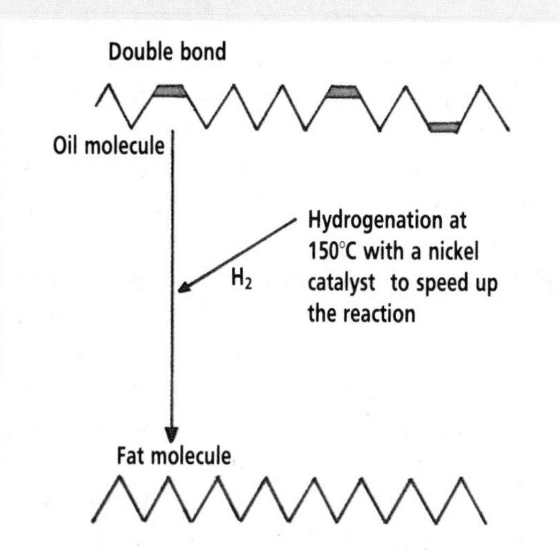

Margarine is a blend of oils, fats and milk. Bacteria in the milk act on some of the compounds to give a buttery taste. Vitamins A and D, which are fat soluble, are added.

Fat molecules are also very long but have no double bonds. They are **saturated** – no more hydrogen can add to their bonds. The fat has a higher melting point than the oil, so the liquid has been converted to a solid.

Things to do

Food for Thought

Things to try out

1 Eggshell is mainly calcium carbonate. Most acids react with the calcium carbonate to give soluble calcium salts.

From the acids you have at home, find out which is best at dissolving eggshell. Here are some ideas to try

- lemon juice
- vinegar
- Coca-Cola.

Things to find out

2 Why are vitamins A and D added to margarine?

ENERGY	735 K/CALORIES 3015 K/JOULES
PROTEIN	0.2 g
CARBOHYDRATE	0.8 g
TOTAL FAT	81.0 g
of which POLYUNSATURATES	41.3 g
SATURATES	14.7 g
ADDED SALT	2.0 g
VITAMINS	% OF THE RECOMMEN DAILY AMOUNT
VITAMIN A	125%
VITAMIN D	300%

MARGARINE IS A USEFUL SOURCE OF ENERGY (CALORIES VITAMIN A (WHICH IS NEEDED FOR GOOD VISION IN DIM AND HEALTHY SKIN) AND VITAMIN D (WHICH IS NEEDED STRONG BONES AND TEETH)

3 Find out which substances found at home people use to put on

(a) bee stings

(b) wasp stings.

Why are different substances used?

4 Survey food labels at home and make a list of the E-numbers of food additives present. The first digit of the E-number usually tells you the type of additive. Use the information in the table to identify the type of food additive in each food.

Type of additive	First digit of the E-number
Colourings	1
Preservatives	2
Emulsifiers and stabilizers	3 or 4
Acids, bases and buffers	5 and some 2/3
Sweeteners/flavour enhancers	4 or 6
Antioxidants	3

Points to discuss

5

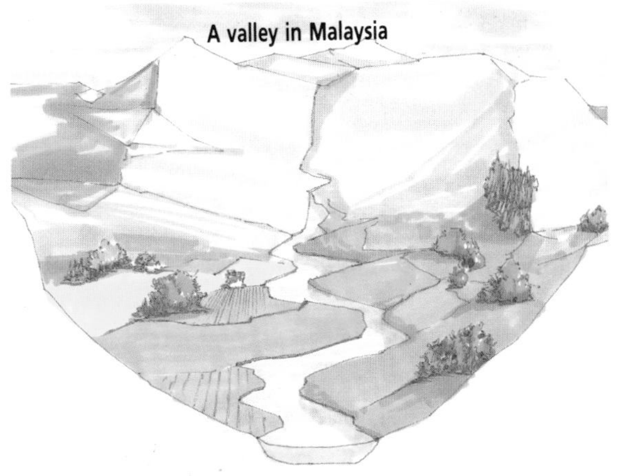
A valley in Malaysia

A dam was built to allow increased irrigation

Once the dam was built a new variety of rice was grown that produced a much better yield of food. It could give two crops a year, and rice output tripled in a few years. Farmers with large farms increased their income by 150%, those with small farms by 50%.

However, the new variety of rice needed more fertilizer than the old one. Fertilizer is expensive. This was no problem for the richer farmers, but those with small farms found after ten years that they were worse off than before the dam was built.

The small farmers sold their land to the farmers who already had large farms. These rich farmers invested in machinery to farm their bigger farms and gradually became even richer. The small farmers moved to the forests of Malaysia and became even poorer.

Is this an acceptable outcome? With the benefit of hindsight, how could the pattern have been changed?

Questions to answer

6 What acids will be needed to make the following salts:

- ammonium nitrate
- ammonium phosphate
- ammonium sulphate?

Write word equations to describe the neutralizations that produce these salts.

Use reference books to write balanced equations for these neutralizations.

7 Use the following words to complete the sentences:

- titrations
- neutralization
- indicators
- fungicides
- phosphorus
- leaching

(a) ____________ occurs when soluble salts are washed out of the soil.

(b) ____________ occurs when equal reacting quantities of an acid and an alkali are mixed together.

(c) ____________-containing compounds are important fertilizers.

(d) ____________ are chemicals which kill certain types of pests.

(e) ____________ can be used to show when an acid has neutralized an alkali.

(f) ____________ are done to work out the quantities of acid and alkali required for neutralizaton.

8 Use the relative atomic masses to work out the relative molecular masses of:

- copper sulphate, $CuSO_4$
- sodium chloride, NaCl
- ammonium chloride, NH_4Cl
- sulphur dioxide, SO_2
- sodium hydroxide, NaOH
- ozone, O_3

9 Using the information on page 85, work out how much ammonia will be needed to

(a) produce 13.2 tonnes of ammonium sulphate

(b) produce 500 tonnes of ammonium sulphate

(c) neutralize 392 tonnes of sulphuric acid.

How much sodium hydroxide would be needed to neutralize 365 tonnes of hydrochloric acid? How much sodium chloride would be produced in this reaction?

10 Iodine reacts with the double bonds in unsaturated fats and oils. The table shows the amount of iodine that will react with 100 grams of a fat or oil.

Fat/oil	Amount of iodine (g)
butter	36
coconut oil	9
lard	56
linseed oil	186
peanut oil	90

(a) Present these figures as a bar chart.

(b) Comment on what the figures tell you about these fats and oils.

11 Ammonium sulphate has the formula $(NH_4)_2SO_4$. Ammonium nitrate has the formula NH_4NO_3. They are both important fertilizers.

(a) Work out the number of nitrogen atoms in each molecule.

(b) Use the relative atomic masses to work out the relative molecular mass of each molecule.

(c) Which of the two fertilizers gives the most nitrogen atoms per tonne?

Relative atomic masses	
Copper	63.5
Sulphur	32
Oxygen	16
Sodium	23
Chlorine	35.5
Nitrogen	14
Hydrogen	1

Index

Scientists mentioned in the text